LO

THE
PUZZLER'S
DILEMMA

THE
PUZZLER'S
DILEMMA

From the Lighthouse of Alexandria to
Monty Hall, a Fresh Look at Classic Conundrums
of Logic, Mathematics, and Life

DERRICK NIEDERMAN

A PERIGEE BOOK

A PERIGEE BOOK
Published by the Penguin Group
Penguin Group (USA) Inc.
375 Hudson Street, New York, New York 10014, USA
Penguin Group (Canada), 90 Eglinton Avenue East, Suite 700, Toronto, Ontario M4P 2Y3, Canada
(a division of Pearson Penguin Canada Inc.)
Penguin Books Ltd., 80 Strand, London WC2R 0RL, England
Penguin Group Ireland, 25 St. Stephen's Green, Dublin 2, Ireland (a division of Penguin Books Ltd.)
Penguin Group (Australia), 250 Camberwell Road, Camberwell, Victoria 3124, Australia
(a division of Pearson Australia Group Pty. Ltd.)
Penguin Books India Pvt. Ltd., 11 Community Centre, Panchsheel Park, New Delhi—110 017, India
Penguin Group (NZ), 67 Apollo Drive, Rosedale, Auckland 0632, New Zealand
(a division of Pearson New Zealand Ltd.)
Penguin Books (South Africa) (Pty.) Ltd., 24 Sturdee Avenue, Rosebank, Johannesburg 2196, South Africa

Penguin Books Ltd., Registered Offices: 80 Strand, London WC2R 0RL, England

While the author has made every effort to provide accurate telephone numbers and Internet addresses at the time of publication, neither the publisher nor the author assumes any responsibility for errors or for changes that occur after publication. Further, the publisher does not have any control over and does not assume any responsibility for author or third-party websites or their content.

First edition: March 2012

Perigee trade paperback ISBN: 978-0-399-53729-5

Library of Congress Cataloging-in-Publication Data

Niederman, Derrick.
    The puzzler's dilemma: from the Lighthouse of Alexandria to Monty Hall, a fresh look at classic conundrums of logic, mathematics, and life / Derrick Niederman. — 1st ed.
        p.   cm.
    ISBN 978-0-399-53729-5 (pbk.)
    1. Mathematical recreations.   2. Logic, Symbolic and mathematical.   I. Title.
    QA95.N483   2012
    511.3—dc23
                    2011042319

PRINTED IN THE UNITED STATES OF AMERICA

10   9   8   7   6   5   4   3   2   1

Most Perigee books are available at special quantity discounts for bulk purchases for sales promotions, premiums, fund-raising, or educational use. Special books, or book excerpts, can also be created to fit specific needs. For details, write: Special Markets, Penguin Group (USA) Inc., 375 Hudson Street, New York, New York 10014.

CONTENTS

ONE

IN THE BEGINNING

The year was 280 BC, plus or minus, and Greek architect Sostratus of Cnidus faced a dilemma that pitted his pride against his life expectancy. The crowning achievement of his career, the Lighthouse of Alexandria, was nearing completion, and Sostratus wanted nothing more than to inscribe his name somewhere on its base. The good news was that there was an abundance of unsullied real estate from which to choose. The lighthouse, destined for superstardom as one of the Seven Wonders of the Ancient World, stood some 400 feet above the island of Pharos in Alexandria Harbor, making it one of the most massive buildings on the planet. The snag was that Ptolemy II Philadelphus, the king of Egypt, had decreed a precise spot for the inscription and insisted that *his* name be the one to occupy it.

Ptolemy II, not unlike Ptolemy I or any Egyptian pharaoh you might happen to think of, lived life by his own set of rules. His moniker of brotherly love derived from his second wife, his full sister. Upon her death, he ordered a deification march in which chariots were led by teams of elephants, lions, leopards, camels, and ostriches, with a bear and rhinoceros thrown in for good effect. His public display of grief was followed by a return to his private concubines. All of which is to say that Ptolemy was a man accustomed to getting his own way, and the penalty for insubordination was high. Nonetheless, as the legend goes, Sostratus was able to chisel for future generations a lengthy inscription that bore his own name. The puzzle is to figure out how he did it without winding up at the bottom of the Mediterranean.

Don't worry. You can keep on reading, because the answer won't be divulged just yet. Where puzzles are concerned, there is nothing worse than being spoon-fed the answer before your wits have been given a fighting chance. Here we will stretch to the opposite extreme, because we're going to go on a tour—a puzzle-solving tour, if you will—with all of its red herrings, false hopes, and frustrations.

Many puzzles, and certainly those crafted from quasi-historical events, must be worded just so. If a puzzle's crafting eliminates all possible ambiguities, the exploration is over before it really starts, a circumstance only slightly less vexing than an accidental glimpse of the answer. At the other extreme, if the wording is too vague, solvers can become distracted by all sorts of oddball theories and won't be able to track a clear path to the solution. For the puzzle at hand, you might seize on the notion that the names of Ptolemy and Sostratus could be connected by an ambigram or some other form of clever typography. Or, in a particularly weak moment, you might concoct an inscription such as "To the great pharaoh Ptolemy II, who commissioned this lighthouse so stratus clouds would not impair navigation." Nice try. Sort of. Lame wordplay is not the stuff of great puzzles, and stratus clouds weren't designated as such until many centuries later. This type of silliness could have been snuffed out right away had the statement of the puzzle included what legend holds to be the actual inscription: *Sostratus, son of Dexiphanes of Cnidus, to the gods who protect those at sea.*

In attempting to solve a classic puzzle, we can either go it alone or use Isaac Newton's technique of standing on the shoulders of giants. Our giant of choice is the late, great Hungarian mathematician George Pólya, long-time professor at Stanford University. Pólya's domains of expertise included probability, combinatorics, and many other branches of higher mathematics, but he spent the latter stages of his 97-year life trying to identify and characterize the essential elements of problem solving. In his seminal work *How to Solve It*, first published in 1957, Pólya gave the following advice to people just starting out on a problem—in other words, people like us. "Do you know a related problem?" asked Pólya. "Here is a problem related to yours and solved before. Could you use it? Could you use its result? Could you use its method? Should you introduce some auxiliary element in order to make its use possible?" Taking the advice of a giant to heart, it remains tempting to think that the Lighthouse of Alexandria puzzle must revolve around some sort of hidden message or quirk of language because subsequent history oozes with such examples. For an example with a special parallel with Sostratus, we fast-forward two millennia to a gentleman named Arthur O'Connor, a prominent figure in the 1798 Irish Rebellion.

The relevant backdrop is that O'Connor, as a member of the United Irishman, had sought French cooperation for asserting the independence of Ireland. Martyrdom dominated his cohort group, but O'Connor was merely arrested and detained at Dublin's newly built Kilmainham Gaol before being transported along with assorted other political prisoners to Fort George in northern Scotland. On his way to this remote new time-share, O'Connor offered a poem as evidence that the prison experience had transformed him into a loyal subject:

> The pomp of Courts and pride of kings
> I prize above all earthly things;
> I love my country, but the King,
> Above all men, his praise I sing.
> The Royal banners are displayed,
> And may success the standard aid.
>
> I fain would banish far from hence
> The "Rights of Man" and "Common Sense."
> Confusion to his odious reign,
> That foe to princes, Thomas Paine.
> Defeat and ruin seize the cause
> Of France, its liberties and laws.

Just what the doctor ordered. Having exalted King George III and denounced American muckraker Thomas Paine in two compact, eloquent stanzas, O'Connor appeared to have created ironclad evidence of his devotion to the king, just the sort of boot-licking encomium that Sostratus was loathe to provide. Except that seeing wasn't quite believing. O'Connor's public choice of rhyme scheme—*aabbcc*, followed by *ddeeff* in the second stanza—wasn't set in stone, if you will, at least not in his own mind. His preferred scheme was *ababcdcdefef*, and when you reorder the lines in that fashion, the poem's eloquence is retained but its loyalism turns to defiance:

> The pomp of Courts and pride of kings
> I fain would banish far from hence.
> I prize above all earthly things
> The "Rights of Man" and "Common Sense."
> I love my country, but the King,
> Confusion to his odious reign.
> Above all men, his praise I sing
> That foe to princes, Thomas Paine.
> The Royal banners are displayed,
> Defeat and ruin seize the cause.
> And may success the standard aid
> Of France, its liberties and laws.

Apparently the only way for O'Connor to express his true feelings was as an inside joke. Not that his ingenious method was the only one available. For a more puzzle-oriented form of poetic reordering, we fast-forward again, this time to Tom and Ray Magliozzi, aka Click and Clack of NPR's *Car Talk*.

> You're four times
> It's hard to
> more likely to
> concentrate on
> have an accident
> two things
> when you're on
> at the same time.
> a cell phone.

In this case the double entendre is unraveled by reading from top to bottom twice, first via the five odd lines and then via the four even lines. In that way you get first statistical and then common-sense support for the Magliozzis' public service campaign that mixing cell phones with driving is as deadly as mixing Leopold with Loeb. But we'd be remiss if we didn't mention that the above creation is printed on a T-shirt that you can win by solving one of *Car Talk*'s regular Puzzlers. And we'd be especially remiss not to mention that one of their November 2005 Puzzlers was a conundrum very much in keeping with our search for hidden messages:

Can you come up with an example of when you might read the English language from bottom to top, so that the next line of text is above the first line?

Great. We were working on one puzzle, and before we've made an ounce of progress we've let ourselves add a different one, two degrees of separation away from the original. But that's not such an uncommon phenomenon along the puzzle trail. In mathematics, it is well chronicled that the 300-year effort to solve Fermat's Last Theorem produced a slew of wrong turns, but many of those diversions led to mathematics that was fascinating in its own right. And in the hierarchy that accountants would call LIFO (last-in, first-out), most of these subsidiary problems were resolved long before the Fermat nut was finally cracked. Somewhat lower on the mathematical tree, I confess that during my fourth year of graduate school, when my PhD thesis consisted of five blank sheets of paper, I deliberately detoured to the goal of constructing a crossword puzzle, being desperate to accomplish *something*. The late Eugene T. Maleska, then crossword editor of the *New York Times*, rejected my first three submissions without breaking a sweat, going so far on rejection number three as to suggest that I wait a long time before submitting another. The fourth try, however, was a charm. It was accepted in late 1980, about a month before I defended my doctoral thesis, which by that time actually had writing on the pages. So I guess I'm a believer in structured digression, if there can be such a thing. It can build confidence.

Except that I took the wrong train on the *Car Talk* Puzzler. I convinced myself that the example of when English might be read from the bottom to the top was—are you ready?—the address on a normal piece of written correspondence.

Don't laugh. Or at least delay your convulsions until I have made my case. Suppose you had been given the task of delivering a letter to me at my old address:

Derrick Niederman

617 South Street

Needham

Massachusetts

If you read the address in inverse fashion, from bottom to top, you get a perfectly logical thought pattern: *Massachusetts*. Sure, but where in Massachusetts? *Needham*. Fair enough. Which street in Needham? *South Street*. I know that one. What number? *617*. Okay, I know the house. *Derrick Niederman*. Oh, that's the name of the guy who lives there? Done. Doesn't that order of operations make sense? Now try the same exercise reading from top to bottom. Here's your new sequence: *Derrick Niederman*. Who's he? *617 South Street*. Where's that? *Needham*. Needham what? *Massachusetts*. Massachusetts? Why didn't you say so in the first place?

The point is not that my answer was necessarily right. It's that as long as I clung to the faint hope that it *might* have some merit, I would never have the mental clarity to divine the actual answer. As it turns out, *Car Talk* was looking for something along the lines of the following:

Zone

School

Down

Slow

That's a message you might find painted on a road, and obviously it has to be presented to passing motorists by inverting the usual order of the words. Just as obviously, it's a better answer than mine.

The sad truth is that when you're trying to solve a puzzle and you're on the wrong train, you have no hope. Trains, by their nature, can take you to faraway places, and in this respect trains of thought are no different from trains of conveyance. Better to have no idea whatsoever than to be on the wrong train. Once you've climbed aboard, the very best thing you can do is find a stopping point at which you can get off, but it's not easy to stop a runaway train. For me, it meant recognizing that my envelope theory was just plain too obscure to win a *Car Talk* T-shirt. And for us, it's time to acknowledge that the whole language/code approach to the hidden message of Sostratus is the very essence of the wrong train. I have a feeling you knew that already. And if we needed any more proof that we were on the wrong train, perhaps it's time to divulge that the actual inscription was apparently written in Greek, as in *ΣΟΣΤΡΑΤΟΣ ΔΕΞΙΦΑΝΟΥ ΚΝΙΔΙΟΣ ΘΕΟΙΣ ΣΩΤΕΡΣΙΝ ΥΠΕΡ ΤΩΝ ΠΛΩΙΖΟΜΕΝΩΝ*. The jig is upsilon.

In a way, glomming on to Arthur O'Connor's inspired but obscure poem of 1798 was unlucky. Our solving process would have been far better served if the meanderings of our mind had chanced on a far more famous poem from the same era, the last six lines of which just might be familiar and in any event read as follows:

> And on the pedestal these words appear:
> "My name is Ozymandias, king of kings:
> Look on my works, ye Mighty, and despair!"
> Nothing beside remains. Round the decay
> Of that colossal wreck, boundless and bare
> The lone and level sands stretch far away.

Percy Bysshe Shelley's "Ozymandias," like O'Connor's jail-house scribbles, features an ambiguous inscription to a king, but its message is altogether different. In lamenting the inevitable erosion of great and powerful civilizations, "Ozymandias" reminds us that our puzzle-solving energies could be devoted not to wordplay, but to the passage of time.

If we are to switch gears in this fashion, let's start by making a couple of basic observations on the subject of time . . . and praying that they lead somewhere. The first is that a modern perspective can be a poor match for a dated puzzle. My initial exposure to this truism was an old mystery story in which the plot hinged around a telephone system, by definition of its time. The central element in unraveling the whodunit was that phone systems of that era didn't allow for an outgoing call until the party on the other end of the prior call had hung up. This wrinkle would have been child's play for a contemporaneous Columbo, but outside the playbook of a modern sleuth.

Inching closer to the methods behind the lighthouse inscription, an engineer friend reminds me that modern industrial prototypes are sometimes fabricated with a 3D printing technique called fused-deposition modeling, in which wafer-thin layers of polymer are laid down in cross sections specified by an electronic file. The intriguing part for our purposes is that not all of these layers are destined for posterity. When the cross sections splay outward into areas where there is no underlying support, the machine is instructed to create temporary pillars made of soluble materials. When the structure has cured to sufficient strength to stand on its own, these pillars are dissolved in sodium hydroxide, leaving everything else intact. Clever, but not available in 280 BC. The only such jettisoning technique available to ancient civilizations was something called "lost wax casting." An item to be cast in bronze was first modeled in wax, and the wax replica served as the negative image for a plaster mold into which liquid bronze was eventually poured. At some point in the journey the original wax model was melted away. Unfortunately, there isn't an obvious path from this labor-intensive process to the answer to our puzzle, but we are reminded that behind an ancient puzzle is an ancient method. Sostratus was resourceful, but he will not blind us with science.

The second observation is a good bit simpler: Planning for the passage of time is fundamental to nature and can be learned by mere mortals. The most common application is allowing for growth. You don't buy snug clothes for your three-year-old. And if you want your garden to feature a dozen lily plants in five years' time, you buy one plant today. Going in the other direction, time can replace hard labor in the removal business. Once kitchens became equipped with dish racks, the phrase "I'll wash, you dry" all but disappeared. Time moves onward and can be harnessed silently, without leaving any traces.

But how much time are we talking about? The original lighthouse puzzle didn't specify, but its wording was more precise than we've had the courtesy to acknowledge. What it did say was "for future generations." That may have sounded like a gratuitous reminder of the permanence of the inscription, as if we were saving a park or a beach "for future generations"—something we have now that we want others to enjoy in the future. But isn't it safe to assume that Sostratus himself never saw the inscription? The puzzle was therefore being literal. His inscription was for future generations and specifically not his own.

This simple observation in hand, we are positioned to give George Pólya his well-deserved second chance. Recall his guidance: "Here is a problem related to yours and solved before. Could you use it? Could you use its result? Could you use its method?" Conjuring up puzzles that depend on the passage of time isn't exactly falling off a log, but now that we are reminded that the protagonist doesn't live to see his handiwork, the following classic pops to mind:

A man is found hanging in a locked room with no furniture and a puddle of water under his feet. What happened?

The answer is that the man was able to hang himself by stepping on a large block of ice, the ice having melted into a puddle of water by the time anyone arrived on the scene. That's the kiddie version, its morbid theme notwithstanding. The adult version, for those adults still playing these types of games, would eliminate the puddle of water altogether, the idea being that more time had elapsed before the grisly discovery.

Sostratus had all the time in the world, so he could afford a slower method. He did not use an obscure technique such as lost wax casting, nor did he build an ice sculpture in Egypt. What he did—and here is your long-awaited spoiler alert—was something in between. Sostratus chiseled his chosen inscription into the base of the lighthouse, all right, but he then covered his words with plaster, into which he carved the obligatory paean to Ptolemy II's fundamental greatness. The unveiling ceremony presumably came and went without a hitch. Over time, however, the plaster could not withstand the elements. In the lexicon of the three little pigs, Ptolemy II got a house made of straw and Sostratus got one made of bricks, and his inscription to himself was the one that withstood the test of time and nature. Unfortunately, in a most unfair instance of "be careful what you ask for," nature outdid itself circa 1300, when an earthquake leveled the Lighthouse of Alexandria, bricks and all. We have to take it on faith that somewhere in between, Sostratus, son of Dexiphanes of Cnidus, had his days in the sun.

I suppose I should apologize again for the wild goose chase that was our solution process. In my defense, I tried to sprinkle a few subliminal hints along the way, stretching to include the words *set in stone*, *plaster*, and *erosion* along our tortuous path. As hints go, though, this is several bread crumbs short of a full Hansel and Gretel, and, even worse, easily lost in the shuffle, just as the Magliozzi brothers would have predicted: It *is* hard to concentrate on two things at the same time.

For an ancient puzzle, the tale of Sostratus had a bit of everything. It created misdirection even though its language was precise. Its solution was aided by knowledge of similar puzzles or structures, but some of that knowledge also led us astray. When we were struggling for an answer, we knew we were struggling, and when we came upon the right line of attack, we knew it was right. We got off the wrong train, but we kept moving. When we let our mind wander into related areas, then repeated the process, we ended up far from our starting point but enriched by our digressions. You can't ask for much more, and we shall see these same themes again as we entertain more modern fare.

TWO

KANGAROO PUZZLES

Back in 2004, I constructed a crossword puzzle whose thematic entries included words such as *contaminates*, *masculine*, *charisma*, *precipitation*, and *unsightly*. The link between these words, not so obvious at first sight, was that they didn't require any clues, at least not in the traditional sense. *New York Times* editor Will Shortz prepared the puzzle for publication by circling certain spaces within the grid, so that the finished entries looked like Figure 2-1.

CON(T)(A)M(I)(N)ATE(S)
(M)(A)S C U(L)I N(E)
(C)(H)(A)R I S(M)A
(P)(R)E C I P I T(A)(T)(I)O(N)
(U)N S I(G)H T(L)(Y)

FIGURE 2-1

If you were able to fill in the circled squares on the basis of the crossing entries, then and only then did the clues pop out: "taints" for "contaminates," "male" for "masculine," and so on. The only thematic entry that warranted a clue was 118A, which read, "Dictionary term for any of the 'self-defining' answers in this puzzle." The answer was "kangaroo word." The dictionary reference was to assure solvers that I hadn't coined the term myself. But there's no reason why you and I can't define a kangaroo puzzle as one that contains its answer somewhere inside.

The diverse nature of kangaroo puzzles suggests that they be presented with laddered degrees of difficulty, just the way it is done in the crossword world. The *New York Times* has long followed a formula of intraweek crescendo that coddles Monday solvers but throws the kitchen sink at whoever might still be standing for Saturday's finale. As for Sundays, constructors are encouraged to take advantage of the extra real estate and produce puzzles that are creative rather than outlandishly difficult. I remember a business meeting in New York years ago, immediately following the publication of one of my first Sunday *Times* puzzles. I passed someone in a hallway who blurted out with a sneer, "I solved your puzzle."

Whoever he was, he continued on his way, pleased at winning the joust but evidently not realizing that most crossword makers *want* people to solve their puzzles. Otherwise we wouldn't go to all that trouble to make the words intersect. As we will soon see, however, the difficulty level of a given puzzle, crossword or otherwise, is often subject to adjustment. The decision of where to place a given challenge within the spectrum of solvability could be considered the most basic puzzler's dilemma.

When it comes to kangaroo puzzles, the easiest forms are the most familiar, and you could say it all started with Groucho. When Groucho Marx hosted the low-budget quiz program *You Bet Your Life* for NBC radio and TV during the 1950s, he spent much of his time ridiculing his guests, but he was also, in a sense, their guardian angel. Faltering contestants would receive softball queries, such as "Who lies buried in Grant's Tomb?" or "What color is an orange?" just to make sure they didn't leave the show empty handed.

A few years after *You Bet Your Life* went off the air, a new show called *Jeopardy!* continued the fine tradition of giving away the answer, and not just because the show gives answers where others give questions. Their version of a kangaroo puzzle isn't as literal as Groucho's, but take a look at some of the freebies they've doled out over the years:

This state, no small potatoes, joins the Union as the 43rd state.

In the early 20th century, Mary Mallon infected about 50 people, not thousands, with this disease.

John Quincy Adams used to skinny-dip in this river that runs through Washington, D.C.

The first question can be answered "What is Idaho?" without knowing that Idaho is the 43rd state. The second question can be answered "What is typhoid?" without knowing that Typhoid Mary's last name was Mallon or that her virulence was exaggerated. And the third question can be answered "What is the Potomac?" without knowing the first thing about John Quincy Adams's bathing habits. Taking these principles one step further, research conducted using the fan-created *J!-Archive* confirmed that *Jeopardy!* contestants should follow their first instincts for a wide variety of question types. In particular, words such as *silversmith*, *cubist*, and *contralto* have achieved category-killer status on the *Jeopardy!* big screen because they are odds-on to point to Paul Revere, Pablo Picasso, and Marian Anderson, respectively. Most often these Pavlovian entries are associated with low-dollar questions, where the show *wants* you to succeed, on the assumption that they'll get their revenge with the higher-dollar amounts.

The kangaroo concept takes a giant leap when you realize that puzzles can emit clues even when no one is actively trying to help us out. Our launching point will be the television series *Murder, She Wrote*, which aired on CBS from 1984 to 1996. The series starred Angela Lansbury as author and part-time sleuth Jessica Fletcher, a latter-day Miss Marple who solved whodunits the old-fashioned way, using intuition, physical evidence, time lines, and other tricks of the gumshoe trade. She talked to suspects. She exposed contradictions. And any clue that was available to her was in theory available to the television audience. But television viewers had an extra sleuthing tool that Jessica didn't, and I'm not talking about the miniature theme from *Jaws* that greeted our ears whenever Jessica spotted an early irregularity. No, the real advantage held by viewers was that any hour-long mystery gives valuable clues by the mere fact that it is an hour-long mystery. The pacing of such a program has to be just so, and in many cases the guilty party can be identified on the basis of timing alone.

If, for example, Virginia Mayo was making a guest appearance, and an incriminating scene involving Mayo's character appeared with 15 minutes left in the program, you could pretty safely cross her *off* the list. *Murder, She Wrote* never spent a quarter of its airtime confirming the guilt of someone already clouded in suspicion. Much more likely was that the murderer would be someone you hadn't seen in a while, someone who reappeared only as the episode was running out of time. That last-second reappearance would inevitably be accompanied by the same sinister background music, this time upped in volume as the ultimate signal of guilt.

A variation of this meta-theme exists in the stock market, in the form of technical analysis. Whereas most investors look for winning stocks on the basis of such factors as a company's earnings, assets, dividend yield, and future prospects, a technical analyst can't be bothered with such details. Why waste your time with balance sheets to identify companies with the potential for favorable share-price movements when you can look at the share-price movements themselves? The basic premise is that certain price patterns are inherently bullish (Microsoft, circa 1989) while other patterns are not (Pets.com, circa 2000). To make money you have to identify the favorable patterns before they turn into, well, unfavorable patterns. In recent years this approach has been applied to shorter and shorter time horizons. The science is inexact and its track record uneven, and I confess that I largely bypassed technical analysis during my years as an investment columnist, but it is a craft that will always have its adherents, and one that can be annoying successful if the market environment happens to produce well-defined uptrends and downtrends.

One such environment was the Internet boom–bust sequence of 1998–2000, during which time I was witness to an extraordinary kangaroo-like inference that I'd like to share. The context is that I was working on what turned out to be my last book-length piece of investment writing. I had been asked by an editor at John Wiley & Sons to write an entry-level investment primer in the form of a murder mystery, and I quickly accepted. My entire mystery-writing experience consisted of 40 short-form mysteries I had written for America Online, so this was something new and different. And it was different on their end as well. Wiley had been strictly a nonfiction publisher for more than a century, specializing in the investment and education markets. Their most recent work of fiction had been written by Herman Melville. My unlikely follow-up was called *A Killing on Wall Street* (what else?), which came out in June 2000, toward the end of the dot-com bubble. During the latter stages of manuscript preparation, I had the strange experience of being asked about a specific stock in a manner that suggested the correct answer all by itself.

The scene was the local Kinko's, where I had gone to make some copies of the page proofs. The pages captured the curiosity of a fellow customer, a disheveled, elderly woman whose woebegone appearance belied her interest in the stock market. She dumbfounded me by asking for my opinion of Celera Genomics, a sky-scraping biotechnology stock of that period. Mindful that doling out free investment advice to strangers has no conceivable upside, I was thankful that I had no official opinion on Celera. The stock had gone up dramatically in the prior two years because of the vast attention and promise then accorded the Human Genome Project. Because Celera was trading on the enticing aroma of a better tomorrow rather than on any established profitability, it was outside the realm of conventional securities analysis. At $200+

per share, the price was frightful by any valuation metric known to man, but to be honest I would have said the same thing when it burst through the $100 barrier just a few months earlier. That's the nature of manias. Technical analysts would have ridden the upswing as long as possible and then cut their losses once it reversed. As long as their system didn't confuse volatility with a genuine sentiment change (the traditional rub for technicians), they could have made out quite nicely.

In the absence of such an approach, the only possible advantage an investor might have with a breakout stock like that would be the ability to track public sentiment, to somehow measure when the genome story had gotten so widespread that 95 percent of all potential buyers were already on board. And now we see where the story is headed. Given that backdrop, my best answer to the woman's question would have been, "In that case, sell," because all I could think of was that if the Celera mania had ensnared a random quirky lady at Kinko's, the bursting of the bubble couldn't be far away. Her question contained its answer. And how. The stock began a swoon that didn't stop until early 2011, 11 miserable stock-market years later, when Celera was bought out by an Indian company called Quest Diagnostics—for $8 per share.

As an investment tale, the Celera story is quite unusual, but when you strip away the particulars, the type of inference it represents is an everyday event for all of us: A phone call whose message is already implied by caller ID, or even the ring of the phone, before the actual caller has said a word. A challenge such as "You'll never guess how old this person is!" immediately suggesting that he or she is significantly older or younger than you would have thought from the photograph in front of you. Hundreds of puzzles have this helpful built-in tell, even ones for which the answer may be out of reach:

A snug-fitting belt is placed around Earth's equator. Suppose you lengthened the belt by one foot and then lifted it from a specific point until all the slack was gone. How high above Earth's surface would that point be?

The full solution to this puzzle involves some delicate algebra and what-not that I'd just as soon stay clear of in a family publication. The answer, however, is merely a number, and it's all too easy to think that the number would be very small, given how minuscule the additional belt length is relative to the 25,000-mile circumference of Earth. So it's something of a shock to hear that the belt would be approximately 121 feet above the ground, until you realize that (1) 121 feet is still puny when compared to 25,000 miles, and (2) the kangaroo element implied a bigger-than-expected result. Even if you didn't have the trigonometric wherewithal to nail down the exact number, a feel for the tenor of the question could have led you in the right direction. Even quantitative puzzles allow for qualitative clues.

In physics, the Heisenberg Uncertainty Principle states that it is impossible to know the position and momentum of a particle with precision. To know an object's velocity we must measure it, and by measuring it we affect it. Even the innocuous step of using light to improve scientific observations can backfire if the light perturbs the delicate quantum particles under consideration. For better or worse, the term *Heisenberg Effect* is applied to a vast array of situations in the social sciences. Perhaps the best known is in anthropology, where so-called participant observations, such as Margaret Mead's groundbreaking investigation of Samoan sex habits, are constantly plagued by the concern that the presence of the observer necessarily affects what is being observed.

When the framing of a puzzle involves stopping the action, a Heisenberg Effect can apply, and the result is an internal clue along kangaroo lines. For example, when a bridge writer asks how a player should defend or declare at a specific stage in the play of the hand, attention is drawn to the key moment as if a gong has sounded, and the success rate of readers will invariably be far higher than it would be at the table. This phenomenon applies with special strength to defenders whose success depends on breaking a standard rule or maxim, such as "third hand high" or "second hand low." These maxims earned their fame by being right most of the time, but the setting of a bridge problem can be a valuable clue that it doesn't work *this* time. Unfortunately, efforts to eliminate this syndrome can result in problems that are too general to convey the point of the exercise.

This same phenomenon can also be looked at the other way around. Consider the chess position shown in Figure 2-2, taken from a tournament game of yesteryear:

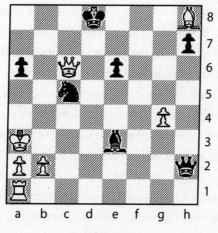

FIGURE 2-2

White is to play, and it's easy to see that if he moves his bishop two squares along the main diagonal, from h8 back to f6, the black king will be checkmated. The position is far too simple to ever be converted into an actual chess problem, even in *Chess for Dummies*. Yet Serbian grandmaster Svetozar Gligorić, whose storied tournament career included four wins over Bobby Fischer, faced this precise position in 1948 and was unable to solve the "mate-in-one" challenge in real time. (Fortunately for him, he still won the game, but not until 17 moves later.)

Just as the person who tells the joke determines the amount of laughter in the room, the person who poses a puzzle can have tremendous sway over its solvability. A puzzle's intrinsic difficulty can be overcome by context, as Puccini's heroine Turandot discovered to her immediate chagrin but ultimate delight. Her third riddle to her potential suitors ("What is like ice, but burns like fire?") was all but unsolvable in the abstract, but her taunting of Prince Calàf led him straight to the answer: Turandot! So if you don't want anyone to solve your clever riddle "What is black and white and re(a)d all over," then don't tell it while clutching a newspaper. If you don't want anyone to figure out what only 12 people have done in the history of mankind, then don't pose that problem at night, outside, under a full moon. And if the newspaper or the man on the moon was what triggered your memory of the puzzle in the first place, understand that those same stimuli are available to your audience. You as the teller can accidentally create a kangaroo element to a puzzle that never had one in the first place.

Moving to our final degree of difficulty, kangaroo characteristics can often be found within counting and sequence puzzles. The following puzzle is a gentle introduction:

**What is represented by the sequence
EOEREXNTEN?**

I'd like to say that I invented this one, and I guess I did in the sense that I thought of it and had never seen it before and got it into print and all that good stuff, but with puzzles you never know. Sometimes you think you've done something truly original only to discover that someone else thought of the same thing 100 years before you were born.

In any event, what I liked about this puzzle was the "TEN" at the end of the letter sequence. It's a reminder that the first step in puzzles such as this is to count how many items you're dealing with, whether they be letters or numbers or anything else. Obviously in this case there are 10 letters, and the answer has everything to do with the number 10 (Figure 2-3).

```
      T           S E
      H F F       E I N
  O T R O I S V G I T
  N W E U V I E H N E
  E O E R E X N T E N
```

FIGURE 2-3

Now for a more sophisticated puzzle with a similar theme:

**What is represented by the sequence
1 1 1 1 1 1 1 1 1 1 1 1 1 1 1 1 1 1 1 1 1 1 3 1 1 1?**

You will have noticed that this puzzle is quite a bit more difficult than "What color is an orange?" There is no answer to be found inside, and no immediately plausible explanation as to why a string of 1s was interrupted in such glaring fashion by a 3. But this puzzle belongs to a large family of puzzles for which enumeration provides a huge clue. If you count up the number of digits in the puzzle and don't get dizzy in the process, you'll get 26, the number of letters in the English alphabet. We're halfway home. All that remains is to recognize that the 23rd letter, W, occupies the "3" slot. Ideally we will sound out the alphabet as we go along, because there's no better reminder that W has three syllables whereas every other letter in the alphabet has but one. The point is that the puzzle could not be posed without including the vital clue, namely the appearance of the number 26.

The principle of recognizing numbers such as 26 for their most significant properties was formally recognized in puzzles of the following form, which debuted in *Games* magazine in the early 1980s:

26 L in the A

12 S of the Z

33 CQ for the IFH

The first one we just solved. The second refers to the 12 signs of the zodiac, while the third, quite a bit tougher, refers to the fact that 33 cars qualify for the Indianapolis 500. Number puzzles of this sort are called ditloids, in honor of the puzzle 1 DITLOID, shorthand for the Solzhenitsyn novel *One Day in the Life of Ivan Denisovich*.

Simple counting can also unearth valuable internal clues for dissection puzzles. Suppose you were asked to divide the shape in Figure 2-4 into two pieces and reassemble those pieces to make a square:

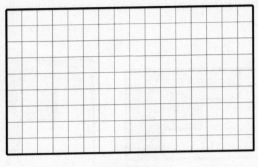

FIGURE 2-4

Before you move another muscle, you have to ask how big the resulting square must be. That's easy: The rectangle measures 9 × 16 = 144 square units, so if it can be transformed into a square, that square has to be 12 units on a side. But you can't just chop off the first four columns of the rectangle, because the two pieces you'd get—rectangles measuring 4 × 9 and 9 × 12—can't possibly be joined to form a square. Instead, the 12-unit sections you create on the top and bottom must be from *opposite* sides, meaning that you'd have to start cutting as in Figure 2-5.

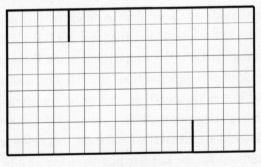

FIGURE 2-5

This modest beginning is of uncertain promise, but if you take it to its logical conclusion, counting all the way, you will produce a stair-step pattern that just might make you believe in magic (Figure 2-6).

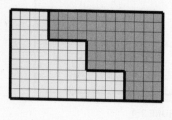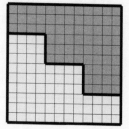

FIGURE 2-6

This zigzag dissection method is not only neat, it's as old as the hills. Italian physician and mathematician Girolamo Cardano described the method in his 1557 opus, *De Rerum Varietate*. Observe that the lattice lines are what gave the puzzle a kangaroo component. The dissection works just as well without them, but finding the solution without a counting mechanism would have upped the difficulty level 10-fold. We would have been forced to think in terms of proportions rather than specific lengths, and on that note it should be mentioned that the zigzag method works only when the sides of the original rectangle are just so. An 8×18 rectangle works because it is of the form $12(2/3) \times 12(3/2)$. The 9×16 rectangle is $12(3/4) \times 12(4/3)$, and so on. There are an infinite number of rectangles that fit the bill, but also an infinite amount of the same area that cannot be transformed into a 12×12 square using this method.

I think you'll agree that if we made the inner clue any smaller than it has been for the last few puzzles, what we'd be left with would no longer be a kangaroo puzzle. It would just be a puzzle, period. We have traveled the entire road. But rather than end the chapter right here, let's close with a few challenges that we can investigate using our new techniques. Could they be made harder? Easier? Take a look at the following four letter-based puzzles and see if any kangaroo tracks turn up.

1. **What does the following sequence represent?**
   ONETWHRFUIVSXGLYDAMBQPC

2. **What do these four words have in common?**
   PRALINES   LIFESPAN
   ESOPHAGI   JACKPOTS

3. **What is the next letter in this sequence?**
   WITNLIT_

4. **What letter can you place on the line, other than E, to complete this sequence?**
   S E Q U E N C _

For 1, the first step is to count the letters, and we see that there are only 23. Three letters are missing, and for the moment we don't care which ones they are. But whatever property is represented in the sequence, three letters in the alphabet don't have it at all.

If we let our eyes wander to the alphabetic maelstrom toward the end of the sequence, we have already bypassed the kangaroo element. The key is that the sequence begins with ONETW, a combination that should look very familiar. If I wanted to give a bigger hint, I'd write the letters down on a Playbill, on the page where the cast is introduced in order of appearance. That's precisely what's going on here. The puzzle presents the letters of the alphabet in order of their appearance in the sequence ONE, TWO, THREE, and so on. Only J, K, and Z are left out entirely. (Sorry, but "zillion" is an expression, not a number.) Of note is that the lead character in the play, A, doesn't show up until the third act—

the first A in the numerical sequence is in ONE THOUSAND. As for the final letter on the list, C holds out all the way to ONE OCTILLION.

For 2, we notice at once that each word has eight letters, so we might as well ask whether that's the common ingredient. The answer is that it's part of the commonality, but unfortunately it's a rather subtle part. We also note that each word contains an A, an S, and a P, a fact that turns out to be relevant but also a bit subtle. The puzzle would be rendered far easier if it followed a query in the same spirit, such as "What do the words MONOPOLY and POLYPHONY have in common?" The answer to that one is that both can be typed out with standard typing techniques using only your right hand. The longest word that can be typed using only one hand is found on the port side—namely STEWARDESSES, but the most delightful entry in this category has to be the compound flower name JOHNNY-JUMP-UP, which is typed with the right hand, hyphens and all.

With the keyboard as our lead-in, we see that when we type the words on our original list, we use each of our eight non-thumbs precisely once, and that's what the four words have in common. There are of course many other words that satisfy this same condition, so it is natural to ask whether any particular choices make for a more difficult puzzle. I would argue that the appearance of PRA-LINES aids the solver's cause, because it calls our attention to the letter P, the one letter that must appear in all words of this type because it is the only letter governed by our right pinky. (Finding words that don't have an A or an S is not easy. The uncommon word PYREXIAL, meaning "to be feverish," avoids the S. To find an A-less word you have to resort to the French word OPTI-QUES.) The choice of ESOPHAGI, on the other hand, camouflages the P marvelously, reminiscent of the puzzles that ask you to count the number of f's in a paragraph. Those paragraphs are inevitably stuffed full of "of"s, which are devilish to count because they don't produce the coveted "f" sound. If you wanted to make the common P more obvious to solvers, you'd throw out ESOPHAGI and use PANELIST instead.

For 3, we're all right as long as we put counting to work in the right way. Once we observe that the number of letters in the question is the same as the number of letters in the sequence (blank included!), we're all set. The letter we're looking for is S, because it's the first letter in "sequence," and each of the other letters got there by being the first letter of the other words in "What is the next letter in this sequence." If we got distracted by the WIT, the LIT, or even by creating NITWIT, that's our problem.

Finally, when I first encountered puzzle number 4, I looked at it a little bit, worried that the answer might be stupid, then asked my significant other whether I could get it. She said I could. About 10 seconds later, I did. The answer wasn't stupid at all, although it was clearly a trick. The puzzle's placement in this section actually might have been rendered even more difficult. Coming after 3, we don't see anything suspicious about the dash at the end of the puzzle, but if we don't direct our suspicions to the dash, we don't have a chance. Yes, the letter E works, as the puzzle tells you, but so does the letter F . . . as long as we place it on top of the dash and convert it to an E in the process!

You may have noticed that the form of give-and-take that enabled me to solve this final puzzle in the first place—finding an opinion that I could, then following through on someone else's optimism—is a powerful kangaroo element in its own right. It's one that we will have occasion to revisit later on.

THREE

THE HUMAN ELEMENT

Puzzles are often used as a diversion from the dreary details of everyday life, but inasmuch as puzzles are man-made and solved by people, it is no surprise that many puzzles sit on a fragile border, with escapism on one side and human nature on the other. The following two tales show what happens when we let puzzles provide some clues about the very human nature we are trying to get away from in the first place. We will then return to our regularly scheduled programming.

REAL-LIFE KNIGHTS AND KNAVES

We spend our lives gauging the reactions of others. When an infant cries, we rush to see what's wrong. When our true love carries a bright smile, we assume that everything is well with the world.

And when we step out to cross a neighborhood street and hear a screeching of brakes, the fist shaking from the driver's seat only confirms our carelessness.

Or does it? Immediate emotional reactions, especially those in the road-rage zone, aren't exactly bound by the axioms of predicate logic. While it's possible that the pedestrian in this scenario was guilty of inattention, contrary details are easy to conjure up. What if the driver was on a cell phone at the time? What if he was speeding? Inebriated? Underaged? Inebriated *and* underaged? What if the incident took place at a crosswalk?

Visitors from the planet Naive would think these scenarios implausible on the grounds that an at-fault driver would display contrition, not anger. Those more entrenched in the ways of Earth know that blame transfer is the number one maneuver from the Freudian playbook. Human beings despise having a spotlight shone on their transgressions, and the speed of denial-based defenses should not be confused with their legitimacy. When guilt-avoidance is at stake, the most ordinary of human minds can operate at Mach 3.

How annoying that guilty and not guilty can produce the same response, and it's not as if this made-up scenario were unique in that regard. If you accuse a man of cheating on his wife, you are certain to provoke outrage, but is it the self-righteous protest of a falsely accused family man or the defensive sputtering of a rightly accused philanderer? More generally, if a transgression—any transgression, really—isn't accompanied by a quick expression of regret, do we conclude that the offender is indifferent to our feelings ("He didn't even apologize!") or in a place 180 degrees away—too mortified to find the appropriate words? Figuring out what evils actually lurk in the hearts of mankind isn't easy. For some assistance we will look at a creation of the delightfully loony logician Raymond Smullyan, a classic logic puzzle known as knights and knaves.

The puzzle has many offshoots, but Smullyan's variations are typically set on an island inhabited by two tribes—the knights, who always tell the truth, and the knaves, who never do. Tourists soon discover the futility of posing questions to the natives. If you come to a fork in the road and ask a nearby islander the way to Sunset Beach, a knight would say left while a knave would say right, leaving you stranded in the middle. The classic knights-and-knaves puzzle, in its simplest form, is to ask the islander one and only one question that gets you where you want to go.

Note the importance of the "one question" part. You can't ask the native a set-up question like "What is the capital of Michigan?" and deduce his knavehood when he answers Ypsilanti. No, that's against the rules. Unfortunately, the direct method—"Say, are you a knight?"—would be even worse, and that's the point of the puzzle. A real knight would truthfully answer yes, but a knave, liar that he is, would also say yes. Logicians call this phenomenon a tautology, which, mathematically speaking, is a statement that is true regardless of the truth values of its parts. The statement "Pterodactyls were either nocturnal or diurnal" fits the mold, as does the more familiar "I may or may not be able to make the party on Saturday night." Both constructs are inarguable and both are utterly unenlightening. To solve the knights-and-knaves puzzle, you must find a way to make the underlying tautology produce some enlightenment after all.

The answer is not exactly obvious, and you should be forewarned that enlightenment was only one of Raymond Smullyan's interests in life. Professionally, Smullyan shed light on subjects as complex as Taoist philosophy and Gödel's Incompleteness Theorem, but when goofing off, his lust for paradox led him to create

masterpieces of obfuscation, including the following fanciful dia-
logue between logicians A and B:

A: Santa Claus exists, if I am not mistaken.

B: Well, of course Santa Claus exists, *if* you are not mistaken.

A: So I was right.

B: Of course!

A: So I am not mistaken.

B: Right.

A: Hence Santa Claus exists!

So if you thought that Smullyan would drop us a hint for his
knights-and-knaves challenge, you are sadly mistaken. But it is
our good fortune that behavioral tautologies of the sort we outlined earlier have not gone unnoticed by the psychotherapeutic
community. The most productive of our hypotheticals turns out to
be the one in which an offender is too mortified to apologize,
which in clinical terms is associated with narcissism. Freud considered narcissism to be an essential part of the human condition
and even admirable in moderation, but its full-time practitioners
find themselves unable to reach out to others, lest they reveal their
own, closely guarded imperfection. A review of the available literature hits an unexpected jackpot in the form of Nancy McWil-
liams and Stanley Lependorf's "Narcissistic Pathology of Everyday
Life: Denial of Remorse and Gratitude" (*Contemporary Psycho-
analysis*, 1990), in which the authors actually invoke the fork-in-
the-road scenario.

When a married couple in which the husband operates
narcissistically reaches a fork in the road on a trip to a new

destination, and is unsure which way to go, the husband will find a way to let his wife pick which road to take. If she turns out to be right, his superior position is protected because he can take credit for letting her choose the way; if she is wrong, he can resent her choice and imply (often nonverbally) that he, had he exercised his own preference, would have gone the other route.

A neat trick indeed, resulting in a win-win for the narcissist and protecting him from the dreaded possibility of being wrong. But it doesn't help us with the puzzle until we recast the husband's approach to his wife as follows: "What would you, as someone other than I, say if asked which direction to take?" This formulation may sound awkward, but it can be translated directly to the time-honored solution to the knights-and-knaves puzzle, which is to ask the native what someone belonging to the *other* tribe would say if asked the way to Sunset Beach. Assuming that the correct answer is in fact to take the road on the left, watch what happens. A knight would respond, "Go to the right" to this convolution, knowing that a knave would lie, while a knave would say the same thing, knowing that the knight would recommend the road to the left but being obliged to lie about it. You simply listen to what the native recommends, go in the opposite direction, and you're now on your way to Sunset Beach, assuming that the sun hasn't already set.

The complication we face on planet Earth is that knights and knaves are not conveniently marked, nor would any such labels be permanent. Dividing the world into good guys and bad guys worked splendidly for Sergeant Joe Friday, but the psychological literature makes it clear that human beings tend to fall from grace in accordance with the situations in front of them rather than the genes inside them. The Routine Activities Theory of the late 1970s, borne of an effort by sociologists to explain an observed uptick in crime even in prosperous times, suggested that most crimes are created by the convergence of three unremarkable factors: sufficient motivation for the potential offender, the availability of a suitable target, and the absence of any guardians. Going a step further, Stanford professor Philip Zimbardo's 2007 book *The Lucifer Effect* offered a stunning account of how ordinary people can commit unspeakable acts. Zimbardo should know: In 1971, he monitored the infamous Stanford Prison Experiment, in which students adopted the roles of inmates and guards in a make-believe prison, only to discover that the role-playing generated a brutality among the guards that forced the entire experiment to be shut down. Even the everyday act of lying has its gradations, with pathological liars outnumbered by the more benign liars.

Our efforts to distinguish liars from truth tellers in the wild usually take one of three related paths. The first path, as risky as it is tempting and fun, is to focus on body language. Someone who looks up and to the right when contemplating a response is deemed suspicious in certain circles on the grounds that up-and-right connotes an effort to dream up something, whereas up-and-left suggests an effort to recall something that actually took place. (Do take the trouble to know whether the person in question is left handed, as the above inferences would be reversed.) Liars are also known to do such things as blink excessively, adopt defensive postures such as folded arms, or even hold an object such as a coffee cup in front of them to create some distance between themselves and those whom they wish to deceive.

Ultimately, any serious effort to interpret body language builds from the knowledge that our muscles can be either voluntary or involuntary. For example, the smile our face forms when compelled by the words *say cheese* uses voluntary muscles, meaning that we have control over them. Involuntary muscles, such as those that emerge in a state of genuine happiness, are different and certainly look more natural if a camera happens to be nearby. The dilation of our pupils is another involuntary mechanism that comes along with surprise or excitement. Psychologist Paul Ekman is credited with identifying these various tells during his many years at the Langley Porter Psychiatric Institute at the University of California, San Francisco. Today, experts in the area of involuntary muscle responses are said to be able to separate liars from truth tellers with an accuracy rate on the order of 85 percent.

The second approach focuses on defensive semantics. Liars are notorious for supplying more detail than needed. Lies are often introduced with qualifying statements, and in the search for formal declarations the liar's use of contractions strangely disappears. Bill Clinton found the qualifier/contraction one-two punch with his historic dodge, "I'm going to say this again: I did not have sexual relations with that woman." Now we know why no one believed him.

The third approach is to consider the intensity of a denial as a contrary indicator, concluding that the stronger the emotional reaction we see from an accused or otherwise defensive party, the more likely it is that the party is *rightly* accused. This line of thinking is quite old. It was already well established by the time Hamlet's mother Queen Gertrude immortalized it with the words, "The lady doth protest too much, methinks," which went on to become one of the most famous, misunderstood, and misquoted lines in all of Shakespeare.

While all three of these lie-detecting approaches are fascinating and unquestionably deserving of more space than was just accorded them, where they fall short is in assuming that we, with our learned insights and unerring judgment, can unravel tautologies all by ourselves. That's a temptation worth resisting. However skilled we may think we are in gauging human reactions, being in the throes of a tautology raises the stakes, because to be wrong means being wrong by 180 degrees. Just ask Othello, who couldn't distinguish his faithful wife, Desdemona, from the adulteress whom Iago wove out of whole cloth. And he certainly couldn't do it via soliloquy. The solution to the knights-and-knaves puzzle offers the unspoken yet powerful insight that you can't escape the clutches of a tautology without taking the time to adopt *someone else's* point of view.

Lest we forget, though, the tourist on the island never did figure out whether he was talking to a knight or a knave. The tourist's real advantage came from the setting of the problem. He, unlike a classically tortured Shakespearean tragic hero, knew in advance that two possibilities existed for whatever responses he received, an advantage we seldom possess in real life. Not only do we have to be alert enough to envision two possibilities instead of one when interpreting someone else's behavior, we occasionally have to do the envisioning from the inside out, when someone else just might be assessing us. Is our shyness mistaken for standoffishness? Does our conviction make us seem closed-minded? And if we enter some brand-new realm and find ourselves clueless or over our heads, do people somehow misread our inexperience and view us as calculating or sleazy instead? Can we somehow break these tautologies in our favor?

Returning to where we started—with the hypothetical rants of an automobile driver of unknown guilt or innocence—it's just vaguely possible that breaking the underlying tautology could reduce the frequency of those rants, aka road rage. Whereas the distance between the participants, coupled with the built-in get-away car, is what makes road rage flourish as a venting mechanism, doesn't that same distance make it impossible to apologize, assuming that's what we want? The natural solution is for drivers to be able to hold up or somehow activate "I'm Sorry" signs. Even confirmed narcissists, for whom apologies are difficult in normal terms, would have a way out. It's worth a try, isn't it? If the experiment succeeded and road rage declined and the world became a better place and all that, the victory would be self-explanatory. And if they didn't work—or, even worse, if "I'm Sorry" signs became as ubiquitous as "Baby on Board" stickers—I'd personally be very sorry, but at least I'd have a little sign I could hold up to say so. That's a win-win.

But sometimes a win-lose is what we should be aiming for. Our very next subject.

FRAMING THE SYMMETRY

A young boy challenges his father to a game of chess. The father, who happens to be a tournament-level player, is pleased by his son's initiative but also wary about setting expectations too high:

"Sure, we can play, but remember, it takes a long time to master such a difficult game."

"I know, Dad," the son says. "I don't expect to win the first time we play . . ."

Then a mischievous grin breaks out.

". . . but if we played two games, there's no way you'd win both of them."

The father doesn't quite know what to make of his son's braggadocio, but the whole set-up sounds like a teachable moment, so he plays along. They bring out two boards and play two games simultaneously. The father quickly sees that his son is right.

What did the son do?

The son starts with the innocent request that he is to play white on one board and black on the other. This seems only fair. White moves first by convention and therefore enjoys a certain advantage. (If you count one point for a win and half a point for a draw, white garnered approximately 55 percent of the available points across all tournament games of the 20th century.) So it makes sense that the father and son should each get a crack at the white pieces. But then the funny business begins.

The father plays white on board 1 and makes his first move. The boy, rather than responding to that move, makes the identical move as white on board 2, then tells the father that it's his move on that board. At that instant the father realizes that he is sunk. No matter how he responds as black on board 2, the son will make that same move with the black pieces on board 1. Obviously the father is simply playing against himself, producing two identical games. In particular, either the son will win a game or both games will be drawn, just as he claimed originally.

The son's coup is known in the game-theory trade as a "strategy-stealing argument." Such arguments have been used to prove that the second player cannot possibly have a winning strategy in games such as tic-tac-toe (which of course will always end in a draw, assuming optimal play). Without reproducing every last detail of the proof, the idea is that if there were such a winning strategy, Player

X could make a random first move and then steal Player O's winning strategy the rest of the way. This works in part because a random first move can't possibly hurt Player X's position—it is never to his disadvantage to have an extra X on the board. (A strategy-stealing argument can't be used for the game of chess itself because of the possibility of the dreaded zugzwang, a circumstance in which any move you make worsens your position.)

Over the years, symmetry arguments have embarrassed game players and game makers alike. In the board game Bridg-It, invented by mathematician David Gale and brought to market in 1960 by Hasbro (when the company was still called Hassenfeld Brothers), each player tries to form a continuous monochromatic path (red or yellow) from one side of the game board to the other. What no one knew back in 1960 is that if Player 1 starts the game by placing a piece of his color in the lower left corner, from that point on he can follow a symmetric strategy that guarantees victory. The good news is that most Bridg-It players were too young to know that the game was strictly determined, as game theorists say when one player or the other has a winning strategy. The same dynamic has saved Connect Four, the popular Milton Bradley game that was solved in favor of Player 1 in 1988 but which still thrives because the winning strategy is well outside the grasp of the grade-school market.

A strategy-stealing argument can be used to calculate the odds of winning a game of Russian roulette. The idea is that a pistol with a bullet in one of its six chambers makes its way to the second player only 5/6 of the time, and when it does, the two players have essentially traded places. The odds of victory for the two players are therefore in the ratio 6 to 5. Given that the probabilities of victory for the two players must add to 1, the likelihood of victory must be 5/11 for the first player and 6/11 for the second. The appearance of fractions is at first unwelcome, but when gamblers talk about 6 to 5 odds, that's precisely what they mean. Russian roulette thereby confirms the fatalistic worldview of Damon Runyon's Sam the Gonoph in "A Nice Price": Said Sam, "I long ago came to the conclusion that all life is 6 to 5 against."

Enough fun and games. The real action starts when we look for strategy-stealing arguments in real life. So suppose that you're still a chess player, say a very good chess player. You frequently get asked for a game by lesser players looking for a chance to maybe take you down a notch. One player in particular—we'll call him Pat, for *patzer*—doesn't seem to take no for an answer. Every time you play—and win—Pat just asks you for a game the following week. You feel you've done your best to discourage him, but none of the subtle clues you give forth are picked up by his primitive antennae. What do you do?

This appears to be a very different problem from the chess puzzle we started with, but it lends itself to the same quest for symmetry. Assuming that you're willing to humor me for a few pages, let's try mimicking the strategy used by the young boy when he challenged his father. To do that, we must identify someone who's measurably better than you are—we'll call him Boris—and pester him with the same force that has been applied to you. In theory, you can't lose. If Boris acquiesces to a series of games, you have an offset for the nuisance on the other end. In the formulation of the original father-son puzzle, that's equivalent to winning one game and losing another. And if Boris turns you down, you take his answer and give it to Pat. That's the equivalent of a draw. Either way, you'll come out 50-50, which is a whole lot better than where you started.

The complication that greets us when we enlarge the scope of the original puzzle in this fashion is that we bring human responses into what had been an emotionless ecosystem. Suppose that Boris turns you down with an abruptness the likes of which you can't imagine transferring to somebody else, no matter how irritating you find him. You then won't be able to follow through on the flip side of the strategy, and your life will officially be asymmetric, though perhaps not for the first time. Part of the problem revolves around not wanting to be the bad guy. In this respect puzzle people have something of an advantage, because they know that bad guys aren't always what they appear:

> A traveler arrives in a small town and decides he wants to get a haircut. According to the manager of the hotel where he's staying, there are only two barbershops in town. The traveler goes to check out both shops. The first one is a mess, and the barber has the worst haircut the traveler has ever seen. The second one is neat and clean; the barber there is perfectly coiffed. Which barbershop does the traveler go to for his haircut?

Obviously the answer is that the traveler goes to the messy barbershop; otherwise there'd be no puzzle. The key step is to recognize that with only two barbershops in town, the two men must cut each other's hair. It therefore makes sense to opt for the barber who does the better job, even if his skill shows up on someone else.

Well, should our lives be symmetric? There's certainly something noble about the concept. It sounds reasonable for a person who doesn't enjoy criticism to be wary of dishing it out, and likewise for hundreds of character attributes or situational preferences.

There is no shortage of aphorisms to spur us on: "People who live in glass houses shouldn't throw stones." "Turnabout is fair play." The ultimate aphorism, and the traditional route to symmetry, is via the Golden Rule: "Do unto others as you would have them do unto you."

Except that the Golden Rule, however noble its intentions, is fatally flawed, because we don't all share the same value system. If, say, one person was raised to enjoy spirited discussions and treats them as pure sport, whereas another was raised to treat argumentation as disrespectful or even hurtful, their joint applications of the Golden Rule have the makings of a train wreck.

Remember, Boris's dismissal of your entreaty may have felt unpleasant, but he wasn't necessarily breaking the Golden Rule. Maybe in his eyes the rejection he handed out was perfectly reasonable. Maybe he's just you, with 10 more years of experience in dealing with unwanted requests. And on the other side of the transaction, it's possible that Pat would have backed off without rancor had you told him firmly that you didn't want to have a regular game with him, even if such an approach was outside your comfort zone. It probably *had* to be outside your comfort zone. By definition, if Pat operated by the same code that you do, he'd have been out of your hair long ago. If we follow that interpretation of the Golden Rule, it's Pat's internal symmetry that matters, not yours.

Some very successful people have thrived on such symmetry. Former Prussian Prime Minister Otto von Bismarck was one. "They treat me like a fox, a cunning fellow of the first rank. But the truth is that with a gentleman I am always a gentleman and a half, and when I have to do with a pirate, I try to be a pirate and a half." Like the boy in the chess puzzle, Bismarck was more of a conduit than an initiator, symmetrizing what was around him through his purposeful role-playing.

Edgar Allan Poe used a similar strategy in "The Purloined Letter," in which Poe's Auguste Dupin solves the mystery of a letter's disappearance by placing himself in the position of the culprit and plotting every move and countermove. Today such posturing is expressed through the expression "It takes a thief."

British social scientist Gregory Bateson, in his essay "The Cybernetics of Self," wrote of symmetric components within arms races and even within the struggles of an alcoholic. Alcoholics Anonymous has long recommended that its followers view alcoholism as a never-ending struggle, and Bateson saw that this prescription, far from an admission of defeat, was a recognition that alcoholics do better by developing symmetric strategies against a constant adversary than through the premature declaration of victory.

Benjamin Franklin is credited with a very different twist on interpersonal symmetry. He was apparently able to preemptively thaw his relationship with an up-and-coming rival in the Pennsylvania legislature without confrontation and even without currying favor in the traditional sense. Instead Franklin wrote to the other gentleman expressing interest in a book in the man's personal library. The upshot was that Franklin borrowed the book for a week or so and returned it with a suitably appreciative note, at which point the relationship between the two men was forever changed for the better. The resulting principle is officially known in social psychology circles as the Ben Franklin Effect. As Franklin wrote in his autobiography, "He that has once done you a Kindness will be more ready to do you another, than he whom you yourself have obliged."

It's nice to hear that some exceptionally smart people have put symmetry to work, but the truth is that most people stink at it. How many times has a football or hockey player retaliated for an opponent's cheap shot with one of his own, only to have the ref see the act of revenge but not its provocation. And how many times has the expression "See how that feels?" been doled out to dismal results, because the response is premeditated and consciously hurtful in a way that the original offense was not. When symmetry is doled out clumsily, the fear of being the bad guy comes to pass after all. Perhaps Emerson was right: "The imitator dooms himself to hopeless mediocrity."

There is in fact a fine line between the individual maintenance of symmetry and the broader problem of transfer, otherwise known as "What goes around comes around." Rather than make up generalizations or conjectures on this fine line in a way that could cost me my armchair psychologist's license, let's consider two stories with contrasting symmetries.

The first one apparently took place on Nantucket or Martha's Vineyard or one of those islands off Massachusetts. The Sunfish—or was it the Sailfish?—was a popular sailing craft of the era, and it seems that it came with a little plug or gasket or something that was essential to its operation. (Your relief at not hearing a psychology treatise from a nonpsychologist has been replaced by the dismay of hearing a sailing story told by a landlubber.) But one day someone set out to have an afternoon sail, only to arrive at her boat and find that piece missing. She did the only reasonable thing . . . stole someone else's. After all, the boats were just lying around, unprotected. And whoever was stolen from on that occasion proceeded to steal from someone else, and so on. Eventually every sailor on the island had a case of gasket paranoia, fulfilling the seaside advisory of British poet Philip Larkin, "Man hands on misery to man. It deepens like a coastal shelf."

A different use of symmetry was chronicled in Josh Leventhal's *Take Me Out to the Ballpark*. The tale was set in the St. Louis area, circa 1900, and the main characters were a pair of next-door neighbors, whom we'll call Al and Bob. Al was building an addition onto his house, and all was going according to plan until Bob advised him in a less-than-neighborly way that he had gone some three feet beyond the property line. The law was unambiguously on Bob's side, so Al retreated. Whether by accident or prescience, Al pulled back three feet *beyond* the property line, setting the stage for revenge at its finest. A few years passed, and wouldn't you know that Bob decided to add on to *his* house. Naturally he planned the addition to be flush with Al's house, consistent with his image of the property line. As the builders put the plan into action, Al saw what was going on but said nothing. He waited, and waited, and waited some more. He eventually pointed out the encroachment, all right, but only after the construction was far beyond the point of no return. A financial settlement was now the only recourse, and Bob ended up giving Al a vacant lot he owned on the other side of town. That vacant lot would eventually be filled by Sportsman's Park, the long-time home of the St. Louis Browns and St. Louis Cardinals. Symmetry can pay. Best of all, no one but Bob would call Al the bad guy.

Obviously the sophisticated revenge of the second story is more pleasing than the juvenile attempt at revenge of the first. A moment's reflection shows us why: The coastal tit-for-tat wasn't even symmetric. The Golden Rule had been applied in its all-too-familiar mutation, "Do unto others as someone else has done unto you." It doesn't work. Even worse, it can't be made into a puzzle.

The real-life analogues to the chess puzzle are only those where you take someone else's precise strategy and throw it back at them. For example, the phrase "Not in my backyard," or NIMBY, became popular in the 1980s in the context of landfills, power plants, and anything else that society as a whole might desperately need but no one wants to see around them. When nuclear power became controversial in the wake of the Three Mile Island accident in 1979, a NIMBY argument was available to environmentalists, who on balance preferred a more natural source of energy, such as solar or wind power. But when wind-power projects finally became feasible, the NIMBY argument was applied to *them* on the grounds that they were eyesores. More common, when a minority party argues against the majority party on an initiative that they might themselves undertake in different circumstances (going to war, raising the debt ceiling, tapping the Strategic Petroleum Reserve), they have to be aware that those same arguments will be used against them as soon as they regain power. When strategy stealing is applied in the public domain, chutzpah and hypocrisy come along for the ride.

Finally, the chess puzzle can be applied to what might be the most important of all forms of symmetry, and that's generational symmetry. When new parents try to focus their attention on the family's newest generation, they cannot help but experience overtones of the generation before. They will face the same challenges their parents did, but this time in the opposite role, and in the best of worlds this change of perspective will bring the generations closer. As a matter of protocol, the telling of the chess puzzle insists on the father exhibiting good humor after he has been hoodwinked. For the son, the true test of symmetry, however symmetric the two chess games were, is whether he is as good natured a generation hence, when one of his own precocious offspring hoodwinks him.

FOUR

TRY LATERAL THINKING

Surely you know this one:

A man walks into a bar and asks for a glass of water. The bartender takes out a gun and points it directly at the man's head. The man says thank you and walks out of the bar. What in the world happened?

Puzzles of this type are best tackled by groups in a question-and-answer session, during which people take turns asking yes-or-no questions of the puzzle poser. This particular challenge is so well known that it's difficult to assemble a group that hasn't heard it, but such is the price of greatness.

One of the features that make this puzzle so memorable is its hidden emotional content. Any and all Q&A solving sessions are doomed to failure until they get around to probing the man's state of mind. The question "Was the man thirsty?" can be a nice start, because the answer "no" usually comes as a mild surprise, but the key query is "Was the man scared?" Well, of course he was scared— he had a gun pointed at his head. Yet the dispassionate facade of the puzzle's storyboard somehow leads solvers astray.

If we are not privy to the innermost thoughts of others, our appraisals of them can misfire rather badly. This sort of thing happens a billion or so times every day but seldom more visibly than in September 1980, when a reporter from CBS Sports conducted an on-court interview with Björn Borg, the legendarily inscrutable Swedish tennis star. Borg was then completing a four-year romp as world number one, and he had just won his semifinal match at the U.S. Open, beating South Africa's Johan Kriek by the improbable tally of 4-6, 4-6, 6-1, 6-1, 6-1. Take another look at that line score. When you focus on the wipeout of the last three sets, and especially when you factor in Borg's absurd Grand Slam winning percentage, his unswerving stoicism, his storied eleventh-hour turnarounds, and everything else that made him the best in the world, it's tempting to believe that the outcome was a foregone conclusion. The interviewer took that bait and asked Borg whether he was worried after losing the first two sets. Borg's reply went something like this: "Was I worried? Of course I was worried! I was down two sets to love in the semifinals of the U.S. Open!" Ever the gentleman, Borg didn't add the words *you fool*, but he didn't have to. The TV audience had already learned the long-concealed truth that Björn Borg was a human being. And so was the nameless bar patron in the puzzle at hand. So why did he express appreciation for being scared out of his wits?

The answer has to do with the visitor's motives. He had sought a glass of water not because he was thirsty but because . . . he wanted to cure a case of the hiccups. The bartender sized up the situation right away and figured that one look at his gun would accomplish the same objective as the glass of water, only faster. And so it was.

The tale of the bartender, the gun, and the hiccups is an example of a lateral-thinking puzzle. The genre has been around forever, originally under the label of "situation" or "yes-no" puzzles, but the process we now call lateral thinking has a very specific champion, who, I'm embarrassed to say, was a stranger to me until 2006. That was the year I joined a quartet of bridge players who ventured from Boston to Verona, Italy, to serve as lambs to the slaughter in an international tournament. Upon arrival at the playing site, we augmented our four-person team with a pair from Malta: Mario Dix and Margaret Parnis-England. Our results, I'm afraid to say, did not exceed our modest expectations, but at some point between sessions Margaret learned that I was a tennis player. She asked me whether I was familiar with the Cyclops line-calling system then in use at Wimbledon.

"Of course," I replied. "What makes you ask?"

"I invented it," she said.

Margaret and I never again talked about bridge. It seems that she dreamed up the conceptual framework of Cyclops and brought it to market with the help of a Maltese engineer named Bill Carlton. As our conversation went on she brought up the name of a fellow Maltese named Edward de Bono, who had coined the term *lateral thinking* in a 1967 book.

As Margaret and then Mario regaled me with de Bono stories, I marveled that the tiny island of Malta, which 10 minutes earlier had been defined by a toy-dog breed and Dashiell Hammett's infamous falcon, was now putting on a world-class display of intellectual firepower. To Edward de Bono, lateral thinking had a very specific meaning. Unlike, say, "Thinking outside of the box," a once-fresh metaphor now dulled by overuse at corporate retreats, lateral thinking means, almost literally, to arrive at solutions by sidestepping the traditional methods of thinking and finding a different path.

Perhaps the best example from Margaret and Mario's cache memory was a tale about an aquarium manufacturer trying to cope with a major problem. Despite the company's best efforts, a frustratingly large percentage of its beautiful, custom-made glass aquariums arrived at their destinations broken if not shattered. Desperate for a remedy that would save their enterprise, the company called on Edward de Bono.

De Bono had no particular expertise in aquariums, glass, packaging materials, overnight freight, or any other form of special delivery. Those skills were as unnecessary to him as cryogenics to the reincarnated. He quickly devised a solution that didn't involve additional costs, insurance, or headaches, and it wasn't to switch delivery companies. His recommendation was simply to write the customer's address on a packing label and affix it to the aquarium without any packing whatsoever. The idea was crazy enough to work, and it did, just as de Bono predicted. Once the handlers could see that they were dealing with glass, the aquariums were treated with tender loving care from pick-up to drop-off. The company's breakage problem all but disappeared.

In lecturing on lateral thinking, de Bono has often used one of the classic puzzles favored by psychologists, the Nine-Dot Puzzle. The challenge is to connect the nine dots in Figure 4-1 by drawing four straight lines without ever taking your pencil off the paper.

FIGURE 4-1

The Nine-Dot Puzzle has been presented in psychological studies as the quintessential thinking-outside-the-box problem, because its solution requires just that—thinking outside of the box created by the dots. The puzzle is widely known, but among first-timers the solution rate is horrendous. Fewer than 10 percent of human guinea pigs manage to solve it.

And, personally, I wonder how many of the psychologists solved it the first time around. I feel reasonably certain that I didn't. My excuse is that I was maybe 10 years old at the time, but I remember conjecturing that the solution involved tilting the lines ever so slightly to enable a line to pass through the left-hand points and, after a long spell, get to a distant locale where the return line, again ever-so slightly angled, would clip the middle points, and similarly for the right-hand side. The flaw in this line of reasoning is that if this particular construction had been legal, it would have required only three lines, not to mention a mile-long sheet of paper. But the actual answer indeed involves extending the line segments beyond the grid, as in Figure 4-2.

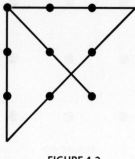

FIGURE 4-2

Although the Nine-Dot Puzzle is extremely well known, it revealed a new dimension during a classroom study in which young students were asked to draw the nine dots on a paper plate. The plates that were handed back showed varying degrees of success, but one particular entry astonished the administrators by displaying a valid solution of a completely different stripe from the one they had in mind. The student drew four lines—horizontal, vertical, and 45 degrees either way—and continued those lines around the back of the plate until they met on the other side, as suggested in Figure 4-3. Given that the whole point of the exercise was to think outside the box, the administrators could hardly award anything less than full credit.

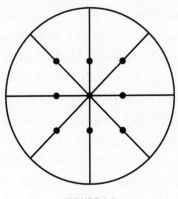

FIGURE 4-3

Jeopardy! questions are famous for their lateral element. Perhaps the best-known example is the Final Jeopardy! question that knocked out 74-time defending champion Ken Jennings: "Most of this firm's 70,000 seasonal white-collar employees work only four months a year." Jennings was immediately sidetracked by the notion that the answer revolved around the holidays, after which there was no recovery. His answer, "What is FedEx?" might as well have had five question marks after it, because he knew it was wrong. "What is Paas?" might have been more inspired, but the huge number of employees, white collar, no less, ruled that out. Meanwhile, challenger Nancy Zerg wrote "What is H&R Block?" on her slate and dethroned the champion. For his part, Jennings received an unexpected consolation prize in the form of free services from both H&R Block and FedEx. As he put it in his online FAQ, "The moral of the story is, if you're going to lose on *Jeopardy!* lose on the corporate question. If your final answer is 'Who was Herodotus?' or 'What is Paraguay?' you're not getting jack as far as endorsements go."

In some sense, lateral thinking has justified its existence by its high-profile applications in the corporate world and beyond. When Peter Ueberroth was tapped to head the 1984 Olympic Games in Los Angeles, he knew that he couldn't follow the blueprint of past financial failures. His solution was to use existing facilities whenever possible (the Los Angeles Memorial Coliseum for the main events, Pauley Pavilion for gymnastics, and so on) and to seek corporate financing otherwise (notably for a swim stadium and a velodrome). In an interview with the *Washington Post* after the conclusion of the games, Ueberroth readily acknowledged that lateral thinking led him away from the sinkhole of public facilities expenditures. The immediate result was a $200+ million profit and *Time* magazine "Man of the Year" honors for Ueberroth. The enduring triumph was that the competition to host the Olympics became something that cities around the world would actually enter and even want to win.

Many of our everyday experiences have the potential to be converted into lateral-thinking puzzles. For example, years ago, when I lived in Newton, Massachusetts, I was scheduled for a morning business trip to New York. Foul weather complicated the logistics somewhat, but the real delay was caused by the sudden disappearance of my two Siberian huskies, leading in turn to my missing the 6:30 train. Given that the house was locked and the only outside access they had was to a fenced-in yard, how did they get away?

This one will never reach the status of the gun and the hiccups, but it has the advantage of being true. It can also be answered via another lateral-thinking puzzle: What gets higher as it falls? The answer is what fell on my house that night. The foul weather that morning was a heavy snowstorm, and the snow from the roof avalanched down onto the fence, shearing one of the panels completely. The dogs had their fun and games for several hours before being found, safe and sound, by someone who lived farther away from me, whether measured in miles or busy suburban streets, than I cared to contemplate. But all was well.

That house seemed prone to strange incidents. There was the time I opened one of my floor-level kitchen cabinets only to see a cat walk out. (I didn't own one, but a neighbor's cat had apparently found a suitable exterior hole and crawled in for the night.) Quite unexpected. And there was the time I arrived home from work and clumsily slammed the back door on my way in. I heard a noise, went to the first-floor bathroom, and found a mirror facedown on the floor, wet on the back. Any ideas? This one was true but fairly preposterous: My inadvertent slamming of the door had created just enough reverberation to dislodge the mirror from its position above the sink. On the way down, the mirror hit the cold water faucet and turned it on. The mirror continued to the floor, *did not break*, but got wet from the stream of water splattering down from the sink above.

The point is that some real-life stories are stranger than fiction and can't be readily converted into first-rate puzzles. No doubt you can do better than my examples, and you might even have fun trying. Some famous lateral-thinking puzzles are merely retooled urban legends, such as this one:

Deep in the forest was found the body of a man who was wearing only swimming trunks, a snorkel, and a face mask. The nearest lake was 8 miles away, and the sea was 100 miles away. How had he died?

The answer is that during a forest fire, a plane had scooped up some water from the lake and dropped the water, swimmer and all, into the forest. The story has long been told as if it were true, but it was officially classified as an urban legend back in 1997 and since then has been further dismantled by the usual watchdogs for such matters, namely Snopes.com and *MythBusters*. The story works fine as a lateral-thinking exercise because the genre tolerates some bending of real-world constraints.

Lateral-thinking puzzles don't get a completely free ride, however, and there is always room for improvement. Case in point is a long-time favorite in which a woman carrying two heavy gold balls tries to escape from a gang of bandits. She weighs just 100 pounds, and each of the gold balls weighs 10 pounds, but, wouldn't you know it, she has to cross a 100-foot wooden bridge over a gorge, and the maximum weight that the bridge will bear is 112 pounds. How does she get across? The answer is that she juggles the two balls as she crosses the bridge, but that's not quite right. Paul Sloane, the author (often with Desmond MacHale) of dozens of volumes of such puzzles, says he has heard from physicists with the smarts to know that the juggling doesn't work. *Myth-Busters* may not debunk this one at an actual ravine, but suffice it to say that the act of juggling produces downward forces that would doom the bridge.

A new and improved version of the bridge puzzle goes something like this:

> **An 18-wheeler is crossing a four-mile bridge that can support only 10,000 kilograms, exactly the weight of the rig. Halfway across the bridge the driver looks on in horror as a one-ounce sparrow lands on the hood of the cab, but the bridge doesn't collapse. Why not?**

Obviously we still have to suspend our disbelief that the weight limit of the bridge can be calibrated so finely. But we also have to take note that whereas the bridge in the juggling puzzle was only 100 feet, this one is considerably longer, and it has to be. The reason the sparrow's landing is safe for the driver is that by driving two miles to the center of the bridge, the truck burns more than one ounce of fuel. This one may be absurd, but it has science on its side.

When he isn't devising new lateral-thinking puzzles or suggesting improvements to old ones, Paul Sloane advises corporations on ways to improve their thinking capacity. The results can be dramatic. Lateral thinking, as applied to corporate branding, can produce adjustments that lead to survival instead of extinction. Sloane notes that a company such as Smith-Corona might have avoided obsolescence by redefining itself as a company that facilitated communication rather than one that made typewriters, thereby substituting a corporate culture that trembled at the thought of technological change with one duty-bound to embrace it. Sloane is also fond of citing the case of media baron Sidney Bernstein, the one-time managing director of a British company called Granada Cinema. In the mid-1950s, the BBC lost its broadcast monopoly, and commercial television stations were being auctioned off throughout the UK. Most observers expected Granada to bid for a license around metropolitan London, where incomes were the highest. Bernstein knew that the licenses for those regions would be expensive, however, so he chose a different path. He instead bought licenses in the Manchester area, north of London, not only because those licenses were cheaper but because the area was *wetter*. In effect, Bernstein viewed his company not as a media conglomerate but as a provider of indoor entertainment, for which persistent rainfall was an obvious match.

Because the definition of *lateral thinking* is so broad, its applications are boundless. It can change corporate history but can also assist our daily lives. The Edward de Bono website features a page with everyday suggestions inspired by his creative team. Nuggets such as the following turned up:

> To avoid forgetting to take things with me when I go out of the house, I throw them next to the front door. When it comes time to leave, I invariably stumble over them on my way out and am therefore reminded of them.
>
> I plant flower seeds in the soil next to gravesites instead of leaving cut or plastic flowers.
>
> I purchase petrol at night when the temperature is lower. As petrol is sold by volume, the product is denser when cooler. Petrol burns by mass, so I get better economy this way.

Where household efficiencies are concerned, there is clearly a fine line between lateral thinking and Hints from Heloise. Yet everyday applications, no matter how simple, are the highlight reel of lateral thinking, and I'll close with a personal episode that was as embarrassing as it was instructive. It arose out of nowhere on my umpteenth trip to the local dry cleaners. For years I had dumped my shirts on the table next to the cash register, counted them up, and announced the total to the attendant. As long as I was the only customer in the store that was fine, but when other people were in line behind me I wished the process would go faster. You'd think that the embarrassing part is how many years it took me to realize that I could count up my shirts *before* I left my house, an epiphany that isn't destined to make it onto the testimonial pages of either Edward de Bono or Paul Sloane, lest their consulting fees plummet. Actually, though, the truly embarrassing part is that I was *proud* of my newfound insight. And perhaps that's the biggest endorsement for lateral thinking. Its potential follows us everywhere, and when we actually get to use it, no matter how trivial the application, we feel a little smarter.

FIVE

KEEPING IT SIMPLE

ometime in the mid-1960s, a Connecticut museum staged an exhibit that included a handcrafted Persian rug, far older than the United States and far bigger than any living room of early New England. Do-it-yourself recorded tours had not yet arrived, so the rug was viewed in groups of 15 or so, overseen by a museum guide. When they reached the wall on which the rug was mounted, the guide announced that the weaving, beautiful as it was, contained a flaw, and he challenged the visitors to find it. This was not one of those two-second flaw searches à la Venus de Milo at the Louvre. What they were looking for was a "Persian flaw." As the guide explained, ancient rug makers were of the view that only Allah was capable of perfection, so they deliberately inserted a mistake to make sure that their god was not offended by their artistry.

Similar folklore in other cultures has produced such concepts as the "humility block" in Amish quilting, the "spirit path" in Navajo rug making, and the "spirit bead" of Native American beadwork. Their common rationale, like their rugs, quilts, and beaded belts, does not withstand close examination. If humans are imperfect by definition, they don't need to worry about producing a perfect piece of art. But the craftsmen weren't willing to take that chance.

What the museum tour was looking for was therefore not the result of weathering or abuse, nor did it come from the sloppiness of a missed stitch or an irregular border. A Persian flaw could be a five-legged horse or other such frivolous abnormality, easily lost in the confusion and passed over at first glance. In one particular group, several glances had come and gone and nothing was turning up. Silence prevailed and capitulation seemed around the corner, but a young girl finally raised her hand and proudly announced that she had found the flaw. The guide played along and asked the little girl if she would tell the others what the rug's flaw was. "Sure," she said. "It's on the wall."

Of course, in the tradition of the proverbial economist who forecasted eight out of the last five recessions, any challenge to identify an error is an open invitation to find fault where none existed. Perhaps the guide added the intro, "Other than the rug not being on the floor . . ." for the next group. But the important point is that the girl saw the problem through her own personal prism, and the result was an effortless simplicity. The battle between simplicity and complexity is a long-standing one in puzzledom, one that will consume us for the remainder of this chapter.

There is in fact an entire family of puzzles that became famous precisely because they offer two routes to the solution, one hard and one easy—or, if you prefer, an adult ticket and a child's ticket, except that when the latter is correct, it is considered far more valuable. One of the classics in this category is a puzzle with a tennis theme:

In the Wimbledon singles championships, the draw consists of 128 players. How many matches are required to produce a champion?

This puzzle has variations only in the sense that Wimbledon needn't be the site and the number of players may be different from 128. But the beauty of this particular format is that it tempts the solver to work backward. Let's see, the final consists of one match, the semifinals two matches, and so on all the way back to the first round, which the problem stipulated involved 128 players and therefore 64 matches. So the total number of matches must be $1 + 2 + 4 + 8 + 16 + 32 + 64 = 127$.

The number 127 has the special property that it can be expressed with its own digits, as in $127 = -1 + 2^7$. Numbers that can be expressed with their own digits are called Friedman numbers, in honor of Erich Friedman of Florida's Stetson University. When the expression places the digits in their proper order, as here, the term "nice Friedman number" applies, and in fact 127 is the first nice Friedman number, as strange as that designation might sound. Math trivia aside, the summation in the previous paragraph expresses a general truth: If you add some number of consecutive powers of two (including $2^0 = 1$), you get one less than the *next* power of two. In particular, if you had 64 players in your draw, you'd need 63 matches, and so on.

You might be smelling a rat right about now. It seems that the number of matches is calculated by taking the number of players and subtracting one, and if that's the case, surely there's an easier way to arrive at the answer. And that's exactly right. Every match in the tournament eliminates one player. Only one player, otherwise known as the champion, doesn't get eliminated at all. The total number of matches is therefore the number of players minus one, and it scarcely matters whether the original number of players was a power of two.

If you found the easy path right away, congratulations. All too often, though, the most elegant approach conks us over the head only after we have clawed out the answer through brute force. That's certainly true of another classic puzzle that offers a simple and a complex solution—the car and the fly.

> **A car travels at 30 mph en route to a destination that is 60 miles away. A fly takes off from the front bumper at the very same time, traveling 60 mph. The fly makes it all the way to the driver's destination, turns around, flies back to the car, turns around, flies back to the destination, and so on. When the car finally finishes its journey, how far has the fly traveled?**

Before we even attempt this problem, we have to get over the fact that a fly couldn't behave this way in real life even if it wanted to. The problem seems to invoke Zeno's Paradox, which basically states that to get from Point A to Point B you have to go halfway first, then from that midpoint to Point B you need to go halfway again, the seeming result being that you'll never arrive at your destination. The notion that an infinite sum can converge to a finite number may have been counterintuitive in Zeno's era (circa 450 BC), but ever since Leibniz and Newton made their discoveries of calculus in the 1600s, the concept of such a limit has had a firm mathematical foundation.

And if you know calculus, you'll have no trouble solving the puzzle, because calculus provides a method of adding things up even when there are an infinite number of them. In this case, the fly travels 60 miles and then turns around, the car having gone halfway in the meantime. When they next meet, the fly will have traveled two-thirds of whatever distance separated it from the car, because it is going twice as fast. The first time this happens that distance is 20 miles, or two-thirds of 30. The next time it happens, it's two-thirds of 20, and so on. When the fly persists, it's up to us to see that the total distance traveled equals the original 60 miles plus 30 times the infinite expression $(2/3 + 4/9 + 8/27 + \ldots)$, a series whose elements are the successive powers of two-thirds. Calculus gives us a way of summing this type of infinite series, and although we won't re-derive the groundwork, suffice it to say that we can plug our various numbers into an established formula and conclude that the fly must have traveled $60 + (30)(\frac{2}{3})(3) = 120$ miles.

Even if you didn't like this method, you have to admit that the answer it eventually produced was nice and clean: The fly traveled precisely twice as far as the car. But wait a minute. Couldn't we have avoided the monstrosity of the prior paragraph by simply noting that (1) the car traveled 60 miles, and (2) the fly was traveling twice as fast over the exact same time period and therefore went twice as far? The 120-mile figure was available to us from the outset, with no assist from Newton or Leibniz or even Zeno. In effect, the streamlined solution results from taking the perspective of the car rather than the fly.

Those of you who are familiar with this puzzle are doubtless aware that at this point in the proceedings it is mandatory to recite the story of the brilliant mathematician and game theorist John von Neumann. The tale begins with von Neumann being given the fly-and-car problem by a student and immediately spitting out the correct answer. Whereupon the student laughed, marveled at his professor's having found the simple path, and said something like, "You know, Professor von Neumann, a lot of people try to solve the puzzle using an infinite series." At which point von Neumann's expression turned quizzical as he replied, "That's how I did it."

The point of this surely apocryphal story was that von Neumann was a very smart fellow, but we didn't need a story to tell us that: In real life, colleagues openly wondered if his existence hinted at a species superior to man, and those colleagues sometimes had Nobel Prizes. For our purposes, the real point is that whatever route von Neumann took to solve the fly-and-car puzzle, it was natural for *him*. For the rest of us, achieving simplicity often means discarding any and all elaborate frameworks to which we might have been exposed. Not so easy, as we will now discover.

In 1981, when strange circular patterns of flattened grain stalks appeared on a hill in Hampshire, England, called Cheesefoot Head, the stage was set for individualized theories. Visitors to the circles reported a wide range of bizarre phenomena, including dramatic changes in barometric pressure, levitation, and the temporary reversal of menopause. For those who pined for evidence of extraterrestrial life and the supernatural, crop circles were a godsend. For scientists, it was an opportunity to explore the possible role of mini-tornadoes called wind vortexes. Other, more speculative theories abounded, including the possibility that the patterns could be traced to run-amok herds of sex-crazed hedgchogs. But those who favored truly simple explanations were vindicated in 1991, when a couple of pranksters named Doug Bower and Dave Chorley could stand it no longer and revealed to the world that the whole thing—make that *most* of the whole thing—was a hoax. They had made the circles all by themselves, using nothing more than some ropes and wooden planks.

In a way, the mystery of the Cheesefoot Head crop circles had been solved six centuries before it ever arose by a farsighted gentleman known as William of Ockham (circa 1285–circa 1345). Ockham is a town in the county of Surrey, England, which borders Hampshire, but the insight was unrelated to local topography. William of Ockham valued simplicity, and his catchphrase of "Entities must not be multiplied beyond necessity" endured as a philosophical concept. The concept was given the name the Principle of Parsimony and, eventually, because it trimmed extraneous thoughts and assumptions, Occam's Razor (using the Latin spelling). The idea is that when different theories vie to explain the same phenomenon, preference should be given to those theories that require the fewest assumptions. On that basis, pranksters versus extraterrestrials is not a close call.

If you were heading the public relations effort for Occam Enterprises, you'd have no shortage of talking points. First on the list would be the razor's frequent application in the realm of medical diagnoses. Attending physicians who follow the principle of diagnostic parsimony are trained to seek a unified explanation when a patient shows up with a laundry list of identifiable symptoms. In the words of Bob Novella of the New England Skeptical Society, "A patient . . . presenting with headache, neck stiffness, fever, and confusion is more likely to have meningitis than to simultaneously have a brain tumor, whiplash, tuberculosis and acute porphyria."

In marketing, the best copywriting is often the simplest. In the 1950s, Lady Clairol revolutionized the hair-color industry by asking "Does she or doesn't she?" and followed up a decade later by asking "Is it true blondes have more fun?" Rome may not have been built in a day, but the Clairol empire was basically built on 12 words. And, according to marketing theorists Jack Trout and Al Ries, a single word is even better, as long as it represents a desirable attribute that the company owns in the public's mind. FedEx means "overnight." Nordstrom means "service," Mercedes means "engineering," and Volvo means "safe."

Simplicity also reigns in politics, at least on the national level, where marketing considerations trump legislative minutiae. Ronald Reagan might not have beaten Jimmy Carter in Trivial Pursuit or even Super Mario Yahtzee, but their electoral battle tipped the other way in part because Reagan's message was simpler and thus better tailored to a sprawling audience.

In criminal investigations, Occam's Razor almost by definition takes a dim view of conspiracy theories, most of which depend on far too many assumptions (or far too many people, with the additional, fatal assumption that they can all keep a secret). On that basis, William of Ockham would have been more comfortable on the Warren Commission than the OJ jury.

In science, simplicity can seem centuries ahead of its time. The theory of continental drift is said to have originated with Flemish cartographer Abraham Ortelius in 1596, as the coastlines of South America and Africa apparently seemed like a perfect fit, even with the primitive maps of that era. But continental drift didn't gain scientific acceptance in the absence of a known force sufficient to move continents, and nothing of that sort was identified until the science of plate tectonics emerged in the 1960s. History repeated itself beginning in 1611, when Johannes Kepler conjectured that the most efficient packing for spheres in three-dimensional space was the standard hexagonal packing. In plain English, Kepler was suggesting that when grocery store owners stack their oranges in the traditional way, they are using the most economical stacking method. This sounds obvious enough, but a formal proof was elusive until Thomas Hales of the University of Michigan put us out of our misery in 1998.

Finally, in language, simplicity is often the best option. We've all heard flight attendants prepare us for takeoff with the supposedly reassuring words, "We'll be in the air momentarily," but we can only hope that they mean "in a moment," not the literal "for a moment" that their choice implies. This rule applies with equal force to the erudite. Early in his speaking career, George Plimpton sometimes thanked his hosts for a "fulsome" introduction as he took to the podium, only to discover that *fulsome* can mean "offensively flattering," "insincere," or even "vulgar." When Plimpton's later speeches used the word *fulsome*, it was only because he was making a joke at his own expense. And William of Ockham would be pleased to know that in the centuries since his passing, society's grammar police have applied Occam's Razor to Occam's Razor itself. We no longer say, "Entities must not be multiplied beyond necessity." Instead we say, "Less is more," or "Keep it simple, stupid."

But the type of simplicity recommended by Occam's Razor is quite specific. In the puzzle world, it is probably best mimicked by the following well-known construction:

Linda is 31 years old, single, outspoken, and very bright. She majored in philosophy. As a student, she was deeply concerned with issues of discrimination and social justice and also participated in anti-nuclear demonstrations.

Which is more probable?
1. **Linda is a bank teller.**
2. **Linda is a bank teller and is active in the feminist movement.**

In a series of well-known experiments conducted by behavioral scientists Daniel Kahneman and Amos Tversky in the early 1980s, 85 percent of the subjects thought that 2 was more probable. The idea, apparently, is that the feminist part resonated a little better, given that 1 doesn't seem very likely based on what we know about Linda. Yet from a probabilistic standpoint, 2 must be even less likely because it represents the conjunction of two events. Logically, it is no different from asking which of the following is more probable:

1. **Tomorrow you will have a sandwich for lunch.**
2. **Tomorrow you will have a sandwich for lunch and meat loaf for dinner.**

The fact that a conjunction of two events is less likely than its individual parts is additional bad news for the already hapless crop-circle scientists. Their slender hope for redemption is that wind vortexes first caused crop circles and that human beings later took over the art, a theory that starts life disadvantaged by having two prongs instead of one.

The trap uncovered by Kahneman and Tversky's experiment goes by the name of the Conjunction Fallacy, and has been widely used to illustrate the nonintuitive nature of conditional probabilities. If the experiment feels a little dated (bank tellers have ceded ground to ATMs, and fewer women are called feminists—or, for that matter, Linda), the good news is that we're getting smarter about this sort of thing. A 2008 paper by Gary Charness, Edi Karni, and Dan Levin revealed that only 45 percent of subjects fell prey to the Conjunction Fallacy. In addition, more heads were better than one. When operating in groups of two, the number fell to 34 percent. Groups of three did even better, erring only 17 percent of the time.

The phenomenon of follow-up experiments not duplicating the results of the original is not new. It even has a name: the Decline Effect. This effect is worrisome in the context of pharmaceutical studies, where the efficacy of new drugs is at stake. But the difficulties underlying the Linda experiment and its ilk have more to do with semantics than a breakdown of the scientific method.

For example, upon seeing the line "Linda is a bank teller and is active in the feminist movement," some people could infer that the simpler line "Linda is a bank teller" is intended to *exclude* involvement with the feminist movement, as if the individual attributes were invitees to a royal wedding, where not being on the list means *not being on the list*. Obviously this misreading makes the conjunction error more understandable. The importance of the adjective *single* in Linda's mini-bio was another surprisingly powerful force. When that one word was inadvertently omitted in one of Charness, Karni, and Levin's follow-up experiments, the error rate was considerably lower.

The moral is that extra information can come with a price. Years ago, I included my middle initial as part of my byline, whether on crossword puzzles or financial articles. That seemed to make sense. Ever since signing my name on a check for the first time (at a hardware store, with my mother looking on), the *C* has been there. But it was soon apparent that the presence of the initial served to confuse rather than clarify. I had to field the question, "What is your middle initial?" surprisingly often, because my choice had created a new variable in people's minds. More was less. A similar issue arises with email addresses for large corporations. From the inside, it's nice to be able to distinguish David A. Green from David B. Green, hence the inclusion of the middle initial in the standard email format. From the outside, however, it can be infuriating not to be able to track someone down for want of their middle name.

Google searches take the concept of extra information to another level and, in the process, shed light on the original Conjunction Fallacy. If we are trying to find our old friend Linda, the stand-alone search phrase "bank teller" probably wouldn't get us anywhere, so we'd add "feminist." In that sense 2 is more likely to lead us to Linda. Sounds good so far, but we have to recognize that we just inverted the Kahneman and Tversky construction. They started with Linda and gave her potential attributes. We're starting with the attributes and wondering if they add up to Linda. One wonders whether some of the subjects in their original experiment unconsciously performed an inversion of this sort.

The preceding paragraph can be viewed in a different but equivalent way. When we add "feminist" to a Google search that already included "Linda" and "bank teller," the probability of isolating Linda might go up, but the number of hits goes *down*. That's how conditional probabilities are expressed in Google arithmetic. And if at any point in our search we are fortunate enough to actually identify Linda, any further attributes we might add run the risk of having the entire search come up empty.

William of Ockham's influence is widespread and deservedly so, but aren't we forgetting something . . . simple? Where puzzles are concerned, there's a limit to the power of simplicity because simplicity needn't be elegant, and great puzzles by definition don't reward inelegant solutions. They're not supposed to be solved by the first method that pops into your head. Three classics stand out for the way in which they mock the straight and narrow. The first is the famous bridge-crossing puzzle:

> **Four people come to a river in the night. The only way across is a narrow bridge that can hold only two people at a time. Because it is nighttime, a lantern has to be used when crossing the bridge. A can cross the bridge in 1 minute, B in 2 minutes, C in 5 minutes, and D in 10 minutes. When two people cross the bridge together, they must move at the slower person's pace. Can they all get across the bridge in 17 minutes?**

This puzzle has the feel of an ancient conundrum, but its first known appearance was in 1981. The simple approach is to take advantage of A's speed and have him carry the lantern throughout and chaperone everyone across the bridge, one by one. The tale of the tape would then read as follows:

STEP	TIME
A takes D across	10
A returns	1
A takes C across	5
A returns	1
A takes B across	2
TOTAL	**19**

That's 19 minutes, so it appears that we came up just short. But the crafters of the puzzle revealed the target of 17 minutes to make sure we wouldn't give up. The puzzle's solution is still simple, but it has a counterintuitive element that takes a bit longer to sort out. Here goes:

STEP	TIME
A takes B across	2
A returns	1
C and D cross	10
B returns	2
A and B cross	2
TOTAL	**17**

We hit our quota, and it's easy to see that the lantern presents no problems—it just gets handed off as needed, and the crossing proceeds. The difficulty of the puzzle is nested in the critical third step, where C and D cross together. The tactic of matching up the slowest participants in this fashion is not unheard of in complex manufacturing processes, where proper synchronization can avoid bottlenecks, but most solvers lack that sort of experience and must reason it out for themselves.

A first cousin to the bridge-crossing problem is the bracelet puzzle. Figure 5-1 shows four sections of a bracelet. It costs 5¢ to cut a link and 8¢ to weld it back together. Can you make a full 12-link bracelet for less than 40¢?

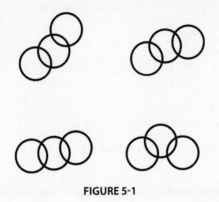

FIGURE 5-1

This one has appeared in various forms over the years, but I've preserved the low cost of the jewelry work to preserve the puzzle's antiquity. At first glance the minimum cost appears to be 52¢. That's the cost you would incur by cutting (say) the rightmost link in each of the four three-link chains, then joining everything into one large bracelet and welding it shut. But, as airlines and delivery companies have figured out, greater efficiency can be achieved by using a central hub. Select one of the three-link chains and cut each of its links. Then use those three links to join the other three three-link chains together, then weld everything shut. That's a total of three cut-and-weld operations, for a total of 39¢.

Because the previous two puzzles were presented with targets, you *knew* that the simplest methods were doomed. Things are made more difficult when the puzzle lacks this tacit assurance. Consider this classic that the legendary British puzzle crafter Henry Dudeney foisted on unsuspecting readers of the *Weekly Dispatch* in 1903 (Figure 5-2).

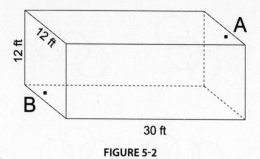

FIGURE 5-2

A spider sits at point A of the rectangular room above, one foot below the ceiling. A fly is located at point B, one foot above the floor. If the spider can only travel by crawling along the walls of the room, what is the shortest distance it must travel to reach the fly?

Simplicity advocates are well aware that the shortest distance between two points is a straight line, so all that remains is the calculation. Let's see. The spider travels 11 feet down to reach the floor, 30 feet to reach the other side, and one foot up to reach the fly. That's 42 feet (Figure 5-3).

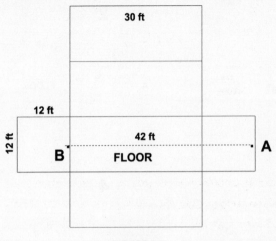

FIGURE 5-3

Yes, that's 42 feet, all right, but it's not the answer Dudeney had in mind. If it were, the spider and the fly wouldn't be much of a puzzle.

In solving the puzzle, the main challenge is to redefine simplicity once we awaken to the cruel reality—with apologies to Douglas Adams—that 42 isn't the answer. The one thing we cannot do is jettison the basic truth that the shortest distance between two points is a straight line. If we abandon simplicity and start to imagine curved paths for the spider, we have no chance at all. We still have to fold the room out as if it were a cardboard box, as we did in Figure 5-4. But we have to acknowledge more choices than were first apparent.

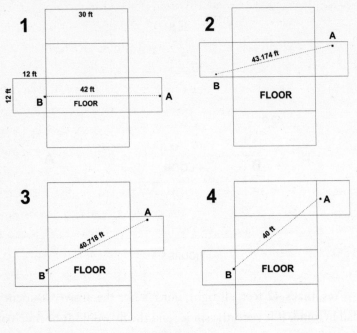

FIGURE 5-4

Fold-out 1 was our initial choice, known to be wrong but at least setting a mark to beat. Fold-out 2 doesn't accomplish that task, but the others do, and fold-out 4 is our winner.

In a 1926 interview in *Popular Electronics*, Dudeney divulged that the idea for the puzzle came to him while sitting in his living room and staring at the ceiling. There was no actual spider traversing the room, but the method of the solution occurred to him immediately, and from there it was only a matter of hammering out the numbers. The elegance of the final answer confirms that those numbers weren't chosen by accident. The desired path is the hypotenuse of a right triangle measuring 24 feet along one leg and 32 feet along the other, which is just the standard 3-4-5 right triangle from geometry class, with each side multiplied by eight. The circuitous path is therefore a full two feet shorter than the original guess. Amazingly, the spider must travel along five of the room's six surfaces to arrive at the destination in optimum fashion.

Similar counterintuitive results arise with airline flight paths. At this writing, the world's longest nonstop flight short of the space shuttle is the 18-hour monster from Newark to Singapore. But if you're wondering whether Singapore Airlines takes you over the Atlantic or the Pacific, the answer is neither. The shortest path, using so-called great-circle navigation, takes you over the *Arctic* Ocean. In that same spirit, Beijing, Shanghai, and Seoul aren't terribly far apart, but if you drew the shortest path from each of those cities to Buenos Aires, almost halfway around the world, you'd go in three completely different directions, cutting across the Atlantic, Indian, and Pacific Oceans, respectively. This divergence arises because somewhere in the middle of the triangle created by the three Asian cities is the antipodal point to Buenos Aires. By definition, all semicircles from Buenos Aires to its antipode are of the same length, so if you make small deviations from that point in different directions, the semicircles you'll meet up with represent wildly different routes.

Puzzles such as the bridge crossing, the bracelet mending, and the spider crawling help set us straight about Occam's Razor. The razor was never intended as a statement that the simplest theory in any given situation will prove correct, even though its message has been hijacked in this fashion countless times. At its core, it is merely a probabilistic guide. All it really says is that the need for extra assumptions can place theories at risk, just as extra parameters can doom a Google search.

When we look back, we see that no item on our prior list of

talking points is immune from attack or counterexample. Continental drift and optimal sphere-packing survived the tests of time and scrutiny, but spontaneous generation did not, nor did the geocentric theory of the universe. In criminal investigations, William of Ockham would have fled the Warren Commission when the canonical single-gunman theory morphed into the hyper-complex magic-bullet theory, but even he would agree that the butler didn't always do it. And diagnostic parsimony is a useful tool but not the law of the land. Hickam's Dictum summarizes the anti-Occam case with a sneer: "Patients can have as many diseases as they damn well please."

Now that the dust has settled, we can close with a compromise. Even if we accept the fallibility of Occam's Razor when examining different theories or options, it still makes sense to look at the simplest ideas first, not because they are certain to work out but because they are more *likely* to work out, and even when they don't, they can be far easier to double-check. No doubt the little girl at the Connecticut museum soon learned that her theory was only second best. What the actual flaw in the rug was has been lost to the passage of time, but to make things current I will admit that the story you just read had a Persian flaw. Did you find it?

SIX

PARABOLIC PARABLES

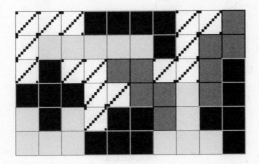

FIGURE 6-1

The 12 pieces that form the patchwork design in Figure 6-1 are called pentominoes. Together they represent all possible shapes that can be formed by linking five squares along their edges, just as Tetris pieces represent all possible four-square shapes. Puzzle fans have been asking all sorts of questions about pentominoes ever since their introduction by Solomon Golomb in 1953. For example, the arrangement in the figure represents one of 2,339 ways in which the pentominoes can fit together to form a 6 × 10 rectangle. And if they can be packed together, they can be spread apart. In the next figure, the same 12 pieces are joined edge by edge, this time to create the biggest possible internal region. Famed computer scientist Donald Knuth is credited with this arrangement as well as the proof that the 128-square area inside the pentominoes of Figure 6-2 cannot be surpassed.

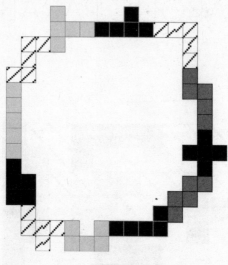

FIGURE 6-2

Pentomino problems are the province of computer scientists and other hard-core puzzle enthusiasts. Although it is obvious that the tightly packed rectangle leaves no internal space, there is no way for us to see at a glance that 128 is the maximum area possible when we spread the pieces out. But questions involving maximum and minimum values are fundamental to certain areas of mathematics.

Suppose you have 100 yards of fencing and are asked to construct a rectangular pen. You can't just make any pen you want, because then there'd be no puzzle. Your pen should enclose as much area as possible. What are the dimensions of this ideal pen?

This puzzle is seldom presented as a puzzle. More often, it is encountered in an introductory calculus course, which is a shame, because you hardly need calculus to figure out what the most spacious enclosure looks like. Having taken some time in the prior chapter to extol the virtues of finding the easy way out, I'm happy to report that the answer to the puzzle can be obtained by a simple combination of intuition and symmetry. Start by laying two 50-yard sections of fence right next to one another, say, in an East-West direction. You have just constructed the worst possible enclosure, one with zero area, as in the left side of Figure 6-3.

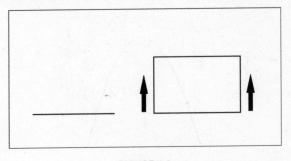

FIGURE 6-3

Now imagine grabbing the fence and lifting it up, as in the right side of the diagram, moving your hands inward as you go along. The area, which started at zero, will of course go up. How long will it continue to go up? As long as a bear can run into the woods—halfway, as the old riddle reminds us, because after that the bear is running *out* of the woods. If you pass the halfway point with the fence, the rectangles you're creating are the same ones you created earlier, now flipped on their sides. The answer to the puzzle, then, is to go halfway, and in this context halfway means creating a square with 25 yards of fencing on a side. I mean, really, what else could it have been?

To give calculus its due, it could have produced the answer even faster, although the one-line calculus solution is deemed unfriendly by some:

$$A(x) = x(50 - x) \Rightarrow A'(x) = 50 - 2x \Rightarrow A'(x) = 0 \text{ when } x = 25.$$

Calculus is equipped to handle a wide range of problems of this type, known in the trade as "max-min" problems. For the fence problem, the sole purpose of the equation in the cryptic one-liner is to locate the highest point in the inverted parabola below, a point that we had already obtained by more rudimentary means.

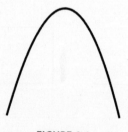

FIGURE 6-4

As you'd expect, calculus can still do its work even when our intuition is found wanting. For example, to add a simple wrinkle to our original problem, we could readily derive the not-so-obvious result that if the pen we were constructing had one side already in place—as in the brick wall of a building—the pen with maximum area would be the one whose length (the side opposite the brick wall) was *twice* its width. Of greater interest, though, are problems of this ilk that for one reason or another elude the clutches of both calculus and puzzledom. This chapter focuses on two challenges that couldn't be more different: One is serious, the other lighthearted; one has been scrutinized the world over, the other not so much. Despite their differences, they are, if you will, *pairable*. Like the fence problems, they both seek to find maximum values, but we must do so without equations and, even worse, without intuition.

■

It has been said that the greatest compliment a mathematician can receive is to be turned into an adjective. An infinite sequence of points that get closer and closer to one another is called a Cauchy Sequence. A one-dimensional complex manifold, whatever image that might conjure up, is called a Riemann Surface. A prime number that remains prime upon doubling and adding one is called a Sophie Germain Prime. And so on. Inevitably, some of these mathematical expressions aren't coined until the honoree has passed away, but a lucky few are around to enjoy the conversion. One of the luckiest of all is economist Arthur Laffer, who found himself adjectivized before the age of 40.

The term *Laffer Curve* first appeared in 1978, in an article written by economist Jude Wanniski for the political journal *The Public Interest*. But the concept was borne several years earlier at a policy meeting at the Two Continents restaurant in Washington, D.C., when the 30-something Laffer took out a cocktail napkin and sketched a diagram that looked something like Figure 6-5.

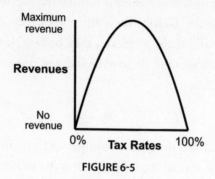

FIGURE 6-5

Controversy has dogged that diagram ever since, starting with a disagreement among the participants as to precisely who was even in attendance. According to Laffer's recollections, the attendees included Donald Rumsfeld, then the White House chief of staff under President Gerald Ford. According to Wanniski, however, Rumsfeld had to cancel at the last minute and instead sent his deputy, a man known to modern readers as former vice president Dick Cheney. But either way we know that the very highest levels of the Republican Party were represented, from which we also know that Laffer hadn't been brought in to talk about fences and square plots, no matter how familiar his little diagram may suddenly appear.

The time was December 1974, near the conclusion of a horrific year for all things Republican. In August, Richard Nixon had resigned the presidency. In September, Nixon was pardoned by his successor, Gerald Ford, immediately making Ford an underdog in his eventual quest for a second term. In October, the new president found himself on the cover of *Time* magazine, but this time alongside the three menacing words of the era: *inflation, recession,* and *oil.* In November, the Democrats gained 49 House seats and

five Senate seats in the midterm elections. By December the Dow
Jones Industrial Average, which had pierced 1,000 toward the end
of the Johnson administration, was struggling to stay above 600.
Even those looking for escapism were out of luck, as *The Brady
Bunch* had been canceled in March.

Still, there was a government to run, and Ford was eager to put
his newly minted WIN (Whip Inflation Now) strategy to work.
The president had announced that he was considering a 5 percent
income tax hike to demonstrate the seriousness with which he
took inflation, then running at the preposterous rate of about 12
percent per year.

The rationale for the tax hike was found in basic Keynesian
thinking: If consumers had a little bit less in their pockets, demand
for goods would subside and prices would stop rising. And if extra
tax revenue could be earmarked for deficit reduction, another
long-term source of inflation would thereby be neutralized. Of
course, it sounds discordant to our modern ear to hear of a Republican president proposing a tax hike, but that's because the supply-
side revolution hadn't happened yet. It began that night at the
Two Continents restaurant.

The concept behind the Laffer Curve is that it is possible for a
government to increase its tax receipts by *lowering* the income
tax rate. The idea sounds paradoxical, and legend has it that Dick
Cheney wasn't an immediate convert, but legend also has it that
the parabolic-looking chart made its napkin debut to win him
over. The leftmost point indicates that if a government taxed its
citizenry at 0 percent, it would receive 0 tax dollars. The rightmost
point indicates that if income taxes were at the opposite extreme
of 100 percent, tax revenues would also plummet to 0 dollars, because the population would lose its incentive to invest, produce,

and earn. If the fencing problem is any sort of role model, somewhere in between should be a magical, if theoretical, place where tax revenues are maximized. In particular, if the prevailing tax rate is to the right of that magical spot, you should indeed be able to increase revenues via a tax reduction. The theory isn't quite as clean as the square-shaped fence, but it hangs together so far.

Wanniski had tried to peddle Laffer's thinking earlier in 1974, but had gone nowhere with the Keynesians in charge at the Treasury Department. The December meeting had the look of an end run around Treasury, but its timing simply reflected Wanniski's hope that the dreadful midterm results would make the White House more amenable to a different strategy. And the truth is that times were different: The inflation that had gripped the economic system had been triggered in part by an exogenous event—namely, the oil spike—a far cry from the garden-variety inflation that stems from an overheated economy. The latter is characterized by low unemployment, but 1974 saw a 3-percentage-point increase in the civilian unemployment rate, from 5.1 percent to a then record high of 8.1 percent. A new word—*stagflation*—had entered the national lexicon, as if to place the Keynesians on alert that a new remedy might be required.

Perhaps Wanniski's quest would have been better served had he invoked the words of John Maynard Keynes himself: "Given sufficient time to gather the fruits, a reduction of taxation will run a better chance than an increase of balancing the budget." For those encountering the Laffer Curve for the first time, Keynes's use of the phrase *given sufficient time* helps unwind the paradox at the heart of Laffer's claims. After all, if the government acted on January 1 of a given year to lower income tax rates for the *prior* year, tax revenues would indeed go down. Economists call that shortfall the "arithmetic effect" of a lower tax rate. But of course

Laffer was referring to the *economic* effect of the tax cut, meaning that the liquidity and incentives created by a tax cut could in theory trigger enough long-term growth in gross domestic product to overcome the near-term arithmetic effect. Here's how those concepts played out at a presidential news conference:

> It is a paradoxical truth that tax rates are too high and tax revenues are too low and the soundest way to raise the revenues in the long run is to cut the rates now. . . . Cutting taxes now is not to incur a budget deficit, but to achieve the more prosperous, expanding economy which can bring a budget surplus.

This quotation might have been among Gerry Ford's greatest hits, but the words weren't his. The news conference in question was held in 1962, when the president was John F. Kennedy and the top marginal income tax rate was 91 percent. Kennedy further explicated just two months before his death:

> A tax cut means higher family income and higher business profits and a balanced federal budget. Every taxpayer and his family will have more money left over after taxes for a new car, a new home, new conveniences, education and investment. Every businessman can keep a higher percentage of his profits in his cash register or put it to work expanding or improving his business, and as the national income grows, the federal government will ultimately end up with more revenues.

Arthur Laffer himself was forever modest about the inspiration that bears his name, aware of the words of JFK and those who came even earlier—far earlier. Laffer not only acknowledged his debt to Keynes but also quoted liberally from the *Muqaddimah*, the magnum opus of 14th-century Muslim polymath Ibn Khaldun: "It should be known that at the beginning of the dynasty, taxation yields a large revenue from small assessments. At the end of the dynasty, taxation yields a small revenue from large assessments." (Even if Laffer's ideas weren't exactly new, it is the nature of mathematical fame that to be an adjective you don't have to be the first person on the scene.) The wrinkle is that the Laffer Curve is a very strange creature, arguably no more a curve than a fake Picasso is a Picasso. For example, you'd be forgiven for glancing at Laffer's napkin doodling and concluding that the ideal income tax rate was 50 percent, in the same sense that the halfway point solved our fencing problem. But nothing could be further than the truth, and it gets worse.

The problem is that the Laffer Curve doesn't really exist, in the sense that its syntax breaks down under cross-examination. What, exactly, would it mean to say that a tax rate of x percent yields tax revenue of y trillion dollars? When do those dollars appear? Next year? The year after? Over five years? And how does the status of the economy figure in to the calculation? When tax cuts are enacted, does it matter if the economy is booming along? Experiencing recession? Suffering from inflation? A single number cannot possibly accommodate all those different scenarios. That's why when Martin Gardner did his own doodling of a Laffer Curve for *Scientific American* in 1981, it came out looking something like Figure 6-6.

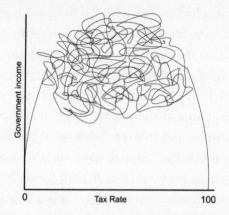

FIGURE 6-6 A REMAKE OF MARTIN GARDNER'S LAFFER CURVE.

Note that the endpoints of Gardner's doodle are as expected, but the mayhem inside is multivalued, the mathematical equivalent of knowing where you are but not knowing where you're going. The whole point of supply-side economics is that great things happen if you're on the right side of the Laffer Curve, but Gardner demonstrated the folly of finding that spot in one stroke of his pen, albeit an extended one.

Gerald Ford never did make the tax cut that Wanniski had sought. Ford's 1975 State of the Union Address, delivered barely a month after the Two Continents meeting, noted that "cutting taxes now is essential if we are to turn the economy around" but quickly cautioned, "unfortunately, it will increase the size of the budget deficit." The political compromise that resulted was a one-time tax rebate, a budget balancer's worst nightmare: The Treasury suffers the arithmetic effect of the cut without an enduring economic effect to compensate. Apparently undaunted, Jimmy Carter tried his hand at rebates right up through his debates with Ronald Reagan, and we know how those went.

Laffer's ideas had to wait until the Reagan administration to gain popular acceptance, but the real tragedy from a fiscal perspective was that the curve was no sooner implemented than it was abused. Reagan budget director David Stockman lamented, perhaps with Caspar Weinberger's military appropriations in mind, that those around him acted as if "additional revenue would start to fall, manna-like, from the heavens." While Keynes talked of "balancing the budget" through lower taxes, presumably he was thinking that spending would remain fairly steady. Not so in practice. While the Reagan financial legacy was a brilliant one in so many ways, given what became of inflation, interest rates, and the stock market during his two terms, the concurrent ballooning of the national debt was a major stain on that legacy. That the political axiom "deficits don't matter" emerged from his political success only made matters worse. The Laffer Curve is often ridiculed for its association with runaway deficits, but that's a bit like criticizing Prozac for the destruction it can wreak when downed with a Harvey Wallbanger.

The Laffer story has a surprising denouement. By the time George W. Bush became president, Ronald Reagan's years in office shimmered with the Day-Glo paint of revisionist history, and the new administration was planning a major tax cut before the inaugural bunting came down. The supply-siders still ruled, but there was a hitch. In selling the Bush tax cuts in February 2001, Federal Reserve chairman Alan Greenspan pitched in by saying that if tax cuts weren't initiated, the budget surplus could grow to the point where the federal government was forced to become an owner of assets. The specter of the United States listed as a shareholder of Home Depot was a frightful thought, but Greenspan's estimates proved to be among the least accurate financial projections in the history of humankind. That, however, isn't the point.

The real point, almost lost in the shuffle, was the Federal Reserve's tacit admission that the United States, with a maximum tax rate of 39.6 percent, was no longer to the right of the Laffer maximum—otherwise, the tax cuts would have been expected to *increase* the budget surplus. That's right. At a time when tax cuts and Republican economic policy had officially been melded into one, the innovation that started it all was thrown under a bus—and nobody noticed.

In 2004, *The Brady Bunch* staged a saccharine reunion, but Jude Wanniski and Dick Cheney did not. Wanniski split with his party over the Iraq War and threw his support behind John Kerry in the general election. Kerry lost, and Wanniski died the following year.

The Water Cube was the swimming venue for the 2008 Summer Olympics in Beijing, and by all accounts it was the fastest swimming pool ever created. No fewer than 30 Olympic records were set at the 2008 games, including all eight of Michael Phelps's gold medal–winning efforts. And you didn't need a Speedo LZR Racer bodysuit to go faster than ever before: The men's 400-meter freestyle and 100-meter butterfly were the only events in which the prior Olympic record *wasn't* broken.

Olympic hosts haven't always cared about the details and technology conducive to superlative swimming times. In 1896, the organizers of the first Modern Olympic Games, held in Athens, balked at the cost of a modern arena and instead pointed swimmers to an existing aquatic venue called the Mediterranean Sea. The Seine River was used for the 1900 Games in Paris, and an artificial lake for the 1904 Games in St. Louis, and it wasn't until the 1908 London Games that swimmers moved inside for good.

But what exactly are the factors that make one pool faster than another? Intuitively we know that cold water tightens our muscles and overly warm water induces lethargy, but competitive swimming venues have long been set between 78 and 80 degrees, eliminating any temperature effect. Pool designers have instead focused on minimizing the turbulence that comes with eight thrashing swimmers in a confined space, and the Water Cube reduced turbulence in several different ways. First, it boasted 10-foot depths, a full meter deeper than prior Olympic venues. A deep pool reduces the vertical wake and is an advantage as long as swimmers can still see the bottom—otherwise they would lose orientation and mistime their turns. Second, the Water Cube used the now familiar tactic of having two extra lanes, making lanes one and eight less of a death sentence for their proximity to a rough border, and amplified this advantage by having water trail off in perforated gutters. The lane dividers, aptly known as "wave eaters," served the same effect. Finally, a superior ventilation system improved the swimmers' breathing by whisking the chlorine smell away.

Former U.S. Olympian Rowdy Gaines, a three-time gold medalist and commentator for the 2008 swimming events, felt that technology had done all it could possibly do to improve swimming times. All that was left, he suggested, was making the water faster, perhaps by changing its chemical composition. But, Gaines said, "I don't know how you make fast water. It's just not possible."

There was no mistaking the grin on Gaines's face, but let's recast his claim in terms of the methodology of square fences and Laffer Curves. The framing is fairly simple. Even if we don't know the first thing about swimming, we know for sure that a swimmer's speed in solid ice is zero. At the other extreme, a swimmer's speed in steam is likewise zero. Somewhere in between must exist a liquid that maximizes speed. Must that maximizing liquid be ordinary pool water? For example, might one swim faster in syrup instead?

The issue at hand is not temperature; it is viscosity, and scientists have been offering opinions in this area for several centuries. Isaac Newton was of the view that the more viscous the liquid, the slower an object's speed through it, and he might therefore have sought a "thinner" form of water. His Dutch contemporary Christiaan Huygens wasn't so sure, leading Newton to hedge his bets in *Principia Mathematica*, the 1687 opus that defined physical laws, derived formulas for planetary motion, and remains unequaled as a contribution to the physical sciences. But it wasn't until the 21st century that a gentleman named Ed Cussler did what neither Newton nor Huygens had the equipment, resourcefulness, or maybe even the permission to do: He filled a swimming pool with syrup and compared the results.

A chemical engineering professor at the University of Minnesota, Cussler understood that a viscous medium created friction and drag, but he wondered in Huygens-like fashion whether swimmers in heavy syrup might benefit by having something tangible to push *against*. Getting the university to yield to his curiosity wasn't easy, but the go-ahead eventually came. In August 2003, Cussler and former student Brian Gettelfinger slowly and methodically filled a 25-meter pool with some 700 pounds of guar gum, a thickening agent normally found in such things as ice cream and shampoo. Guar gum is edible but not particularly appetizing in large quantities, so much so that Cussler felt obliged to take the first practice run in a pool that now approximated the world's largest bowl of mucus.

Once Cussler pronounced the pool fit for humans, a special race was on. A volunteer group of 16 competitive and recreational swimmers swam an assortment of strokes in both the special pool and an ordinary one (Gettelfinger himself was a near Olympian but resisted the urge to participate). The results of the experiment were stunning to everyone but Cussler. The viscosity of the water created no measurable difference between the swimmers' times.

It would be tempting to go a step further and conclude that the guar gum pools were even faster than their clear counterparts, on the grounds that swimming in something that disgusting would have to take a fraction of a second off a swimmer's time. But Cussler and Gettelfinger showed their professionalism by sticking with the numbers and were rewarded with the 2005 Ig Nobel Prize in Chemistry. The Ig Nobels are awarded annually at Sanders Theater at Harvard University by a publication called the *Annals of Improbable Research*. The categories follow the mold of actual Nobel Prizes, but the winners do not. Whereas the 2005 Nobel Prize in Physiology or Medicine was won by Australia's Barry Marshall and Robin Warren, who demonstrated the role of the bacterium *Helicobacter pylori* in the formation of peptic ulcers, the 2005 Ig Nobel Prize in Medicine was won by Gregg Miller of Oak Grove, Missouri, who invented artificial testicles for dogs. What made the 2005 Ig Nobel choices especially interesting is that Cussler and Gettelfinger were not nominated for the prize in fluid dynamics, their seemingly natural category. To have won in that category would have required beating out Victor Benno Meyer-Rochow and Jozsef Gal, who took home the prize by calculating the force with which penguins defecate.

Left open for future researchers is the determination of the precise shape of the viscosity/speed curve for swimming. There are at least two possibilities consistent with Cussler and Gettelfinger's research. One is that a swimmer's speed remains fairly constant throughout different viscosities, but then falls rapidly when the conditions become too extreme. The speed chart might look something like that shown in Figure 6-7.

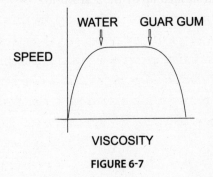

FIGURE 6-7

But then there's also the possibility that somewhere in between ordinary pool water and guar gum, there is a viscosity that generates the greatest speeds of all. Graphically, that case would have the shape shown in Figure 6-8.

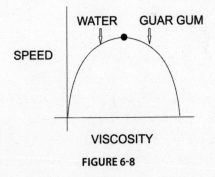

FIGURE 6-8

In guessing what happens between pool water and guar gum, we're back to where we started. Knowing what happens at the endpoints of a problem tells us frustratingly little about what happens in between. Calculus problems have a reputation for gnarliness that I don't wish to spoil, and pentomino problems have barriers to entry all their own, but neither is a match for the maxmin problems drummed up by the real world.

SEVEN

LITTLE BIG JUMP

When I was all of three years old, just the right age to absorb bedtime stories, the Beginner Books imprint at Random House, fresh from its successful launch with Dr. Seuss's *Cat in the Hat*, published "The Big Jump," by Benjamin Elkin. The story revolved around a young boy named Ben, who lived in a kingdom with a strange set of rules. The rule introduced on the first page was that the only person in the kingdom who could own a dog was, you guessed it, the king. But one day, when one of the king's puppies took a liking to Ben, that rule began to bend. The king offered Ben the opportunity to own the puppy if he could jump from the ground to the top of his castle, and proceeded to demonstrate that gravity-defying maneuver. To make a short story only somewhat shorter, Ben went home and practiced his jumping to no real avail. He could jump two boxes' worth by himself and another two with the help of a pole, but no more.

When he returned to the castle, he used a trick that the puppy had demonstrated while frolicking alongside his four-box pile. Ben jumped to the top of the castle, all right, but he did so without magic. He simply jumped one step at a time all the way up the stone staircase that conveniently surrounded the castle, from ground level to the uppermost turret. The puppy was his reward.

I confess that as a child I found Ben's solution to be a bit cheesy. The transition from the king's actual big jump to his acceptance of Ben's serial jumping was facilitated by some linguistic sleight-of-hand, including the king's after-the-fact claim (using a dozen of the 369 mostly monosyllabic words preapproved for Beginner Books), "I did not say that Ben had to do it in *one* jump." As far as I was concerned, what Ben had done was to replace a task he couldn't perform with one that he could, at which point he received a very favorable ruling.

In certain circles, substitution is permitted when your first choice is unattainable. In *Finian's Rainbow*, Og the Leprechaun transitioned his affections from Sharon McLonergan to Susan the Silent with "When I'm Not Near the Girl I Love (I Love the Girl I'm Near)," laying the groundwork for Stephen Stills to extend the thought a generation later with "Love the One You're With." Less exotically, the question-and-answer fly swatting that passes for presidential debates has devolved to the point where both parties follow the advice of former defense secretary Robert McNamara: "Never answer the question that is asked of you. Answer the question that you wish had been asked of you."

Can we really do that? Harvard Business School professors Todd Rogers and Michael Norton explored the matter in their 2010 working paper "The Artful Dodger: Answering the Wrong Question the Right Way." One of their mildly depressing conclusions was that speakers can achieve higher ratings by answering a similar question fluently rather than making any sort of hash of the original question. Those who dodge questions don't necessarily pay a price because their listeners are busy making their evaluations by a range of criteria, by no means restricted to the questions and answers in front of them.

Puzzles are a wee bit different. When presented with a puzzle to solve, we can't very well substitute a different one that's more to our liking. What we *can* do, though, is appreciate that the insight obtained from well-chosen substitutes can potentially help us solve the original.

The idea of solving a problem via substitution has applications across all difficulty spectrums. Recall that when Andrew Wiles proved Fermat's Last Theorem—demonstrating that the equation $x^n + y^n = z^n$ has no solutions in positive integers when $n > 2$—he didn't write about Fermat's Last Theorem per se. Instead, building on prior advances by Gerhard Frey and Ken Ribet, he proved a form of the Taniyama-Shimura Conjecture, which held that every elliptic curve has a modular form with the same Dirichlet L-Series. You see the problem here. The reason Fermat's Last Theorem gained such popular acclaim is that its original statement was familiar and comprehensible; it was merely the Pythagorean Theorem with a change of exponent. But what in the world is an elliptic curve? A modular form? A Dirichlet L-Series? What Wiles did worked for him but was unrecognizable as a strategy for the rest of us: We want to replace difficult problems with equivalent problems—so-called isomorphs—that are *easier*. The oil-and-vinegar puzzle, something of a classic but hardly overused, provides us with a launching pad for this important technique.

Suppose you have 20 quarts of oil in one container and 20 quarts of vinegar in another. You transfer five quarts from the

oil container into the vinegar container, at which point you mix
those contents up as best you can. You then take five quarts of
that mixture and return them to the oil container. The question,
which you have surely anticipated, is whether there is more oil
in the vinegar or more vinegar in the oil.

This puzzle is a bit of a fooler. It is tempting to believe that
because the liquid originally taken from the oil container was pure
oil, whereas the amount transferred back is a combination, that
there is more oil in the vinegar than vinegar in the oil. A little
algebra would set us straight, but we'd like to arrive at a solution
with as few tools as possible.

The key step is to approach the problem via discrete units
rather than amorphous blobs. To wit, suppose you have a pile of
20 white discs and a pile of 20 black discs. (Imagine, for example,
the pieces in the ancient Chinese game of Go.) You take five of
the white discs and place them into the black pile, which you give
a thorough mixing. You then return five of the discs from that
mixture (imagine yourself blindfolded, if you like) into the original white pile (Figure 7-1). Do you have more black in the white
or white in the black?

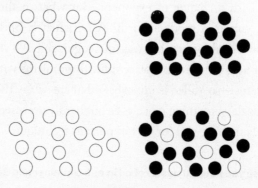

FIGURE 7-1

Somehow the restated problem is far more accessible than the original, even though the algebraic solutions to the two problems would be identical. But we're not quite done with our simplifying. Forget that I just asked what happens when we transfer five discs one way and another five the other way. What happens if you choose just one disc? Now *that* we can handle. Take a single white disc from the left to the right, and then take a single disc, arbitrarily chosen, from right to left. If the second disc happens to be the white one, all you've done is given that disc a round trip back to the starting position, changing absolutely nothing. And if the second disc is black, all you have done is to make a switch, changing the *totals* of black and white in each pile but not the balance. In terms of the original puzzle, the amount of oil in the vinegar is therefore the same as the amount of vinegar in the oil. Yes, we're done, and yes, the problem is that easy. It just seemed difficult because the oil/vinegar visual gave us nothing to count. And if there's any lingering doubt that the simplified puzzle is equivalent to the original, just imagine that a white disc is an oil molecule and a black disc is a vinegar molecule, and use as many discs as you need. Everything checks out just fine.

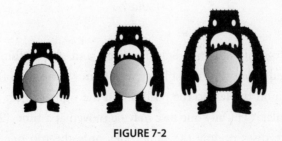

FIGURE 7-2

Substitution can provide clarity to a puzzle that looks absolutely hopeless. Figure 7-2 depicts a puzzle featuring three monsters of different heights—small, medium, and large. Each monster carries a small globe, but medium and large sizes are also available. At any given moment, it is possible for any monster to change the size of his globe, subject to three rules of monster etiquette: (1) Only one globe may be changed at a time, (2) if two globes are the same size, only the globe carried by the bigger monster may change, and (3) a monster may not change the size of his globe to a size carried by a bigger monster. Your objective is to create a series of steps that ends with each monster carrying a large globe. Or not. This puzzle sounds like a complete mess, and it is of little consolation that the challenge can be rendered far more intelligible and inviting when converted to an electronic format, in which noises and prompts can advise you if the moves you make are in accordance with the rules. Not having an applet handy, we will instead use the diagram in Figure 7-3, which may look familiar in its own right.

FIGURE 7-3

Instead of thinking about monsters and globes, think about discs and poles. Our new objective is to move the discs from the position on the left to the position on the right, subject to these three rules: (1) Only one disc may be moved at a time, (2) if two or more discs occupy the same pole, only the top one may be moved, and (3) at no point can a disc be placed on a smaller disc. Just like that, the monsters and globes puzzle has become the

Tower of Hanoi, the official name for the discs and poles puzzle. In the process, it has become much easier to solve.

In real life it was the other way around. The Tower of Hanoi puzzle was invented by French mathematician Édouard Lucas in 1883, so it's been around a lot longer than the monsters and globes puzzle, applet or none. The monsters and globes challenge was literally a laboratory creation, brought to life by psychologists interested in comparing the solving times for the original Tower of Hanoi with those for other, logically equivalent puzzles (known in the trade as "representations"). What makes the monsters and globes representation especially unfriendly is that the constraints are not intuitive. In the absence of electronic backup, solvers must always double-check to see whether a given move is legal. By contrast, the Tower of Hanoi puzzle gives us visual feedback, and rule 2 is superfluous, because it's obviously forbidden to move a disc that isn't the highest piece of whatever pole it happens to occupy. (Figure 7-4 vivifies the connection and shows how the moves of the two puzzles correspond to one another.) The point is that if history had been inverted and we had been saddled with the monsters and globes puzzle, we would have been well advised to turn to the Tower of Hanoi as a substitute. Under laboratory conditions, the average solving time for the monsters and globes puzzle was about 30 minutes. The Tower of Hanoi puzzle took just 2.

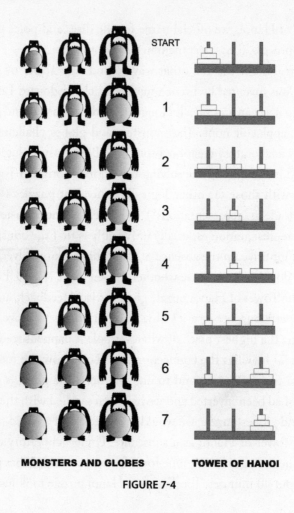

MONSTERS AND GLOBES   **TOWER OF HANOI**

FIGURE 7-4

Even simple tasks can be aided by the right representation. Years ago I spent a lot of time tutoring high school mathematics students in my area, and I watched all too many of them struggle with the challenge of multiplying 25 times 16. The first snag was that students would instinctively reach for their calculators, on the grounds that 25×16 is superficially indistinguishable from, say, 37×17, where no convenient shortcut is available. But the whole point of the exercise is that 25 is a very friendly number within the decimal system, so that the problem becomes easy if you think of it as a matter of sixteen 25s as opposed to twenty-five 16s. Some students will still balk, and for those students there is a follow-up question: "How much money do you have if you have 16 quarters?" This recasting makes the appropriate grouping obvious, and even the weakest, most calculator-dependent students produced the answer of $4.00. Now and only now can you get rid of the dollar sign and the decimal point to produce your final answer. Money has relevance that abstract equations do not.

When we can't literally substitute one puzzle for another, we can try extending the methods of absurdly simple puzzles to meet more challenging tasks. Suppose you encountered a problem like the following:

Choose 20 numbers from the set {1, 4, 7, . . . 97, 100}. Show that there must be two distinct numbers whose sum is 104.

No fun, you say. Too hard, you say. Well, you have a point. The problem is taken from the 1978 William Lowell Putnam Examination. The Putnam exam is a mathematics competition administered each December by colleges across the United States, and for most students it is the hardest test they will ever take. The exam consists of six problems in the morning and, for those students brave enough to return, another six in the afternoon. At 10 points per question, a total of 120 points is theoretically available, but the problems are baffling, and the committee is notoriously stingy when it comes to awarding partial credit. In 2006, 3,600 of America's best and brightest math students took the exam, and 2,260 of them got no points whatsoever, one of five occasions to date when the median score was 0. In 1978, the median score on the exam was 11. Getting this one problem right would put us within a point of the median, and we'll start the solving process with a substitution that should provide a nice ego boost:

> A drawer contains 8 black socks and 6 brown socks. It's dark in the room, so you can't see inside the drawer. What is the smallest number of socks you'd have to pull out to be guaranteed a matching pair?

Ah, that's more like it. This one we saw in grade school. We got it right then, and we'll get it right now. There are two different colors of socks. All we have to do is choose three socks and we're guaranteed to have a matching pair. We don't know which color the socks will be, but two of them are guaranteed to match. The "8" and "6" in the puzzle are pretty much irrelevant, because as long as you have at least two of each color, everything depends on the number of colors, not the number of socks. If we had three different colors we'd need to choose four socks, and so on.

You might be thinking that the Putnam problem didn't have socks. Or did it? From the original set {1, 4, 7, . . . 100} we pluck the following 16 pairs:

{4, 100} {7, 97} {10, 94} {13, 91} {16, 88} {19, 85} {22, 82} {25, 79}

{28, 76} {31, 73} {34, 70} {37, 67} {40, 64} {43, 61} {46, 58} {49, 55}

There are your socks. Each pair of numbers adds up to 104, the target established by the problem. The only numbers left out are 1, because its would-be mate of 103 wasn't on the list, and 52, because it would need another copy of itself to reach 104.

The identification of these pairs harkens back to what the great

German mathematician Carl Friedrich Gauss is alleged to have figured out as a schoolchild, when his teacher made him add the numbers 1 through 100 as a means of occupying his precocity. Gauss responded by matching 1 with 100, 2 with 99, and so on, all the way to 50 and 51, thereby creating 50 pairs that each added up to 101. The required sum of 5,050 was presented within minutes to the now exasperated teacher, who doubtless wished he had asked young Gauss to work on Kepler's sphere-packing conjecture instead.

Now let's tinker with the original wording and ask the following question: How many numbers would you have to choose from the set {1, 4, 7, . . . 97, 100} to be certain of finding a matching pair—er, that is, a pair that adds up to 104? Well, if you were really unlucky you'd choose 1, 52, and precisely one number from each of the 16 sets. That's 18 numbers. When you choose your 19th, you would be guaranteed to have chosen both members of one of the pairs listed earlier, and by definition any such pair adds up to 104. The Putnam problem, in a rare moment of generosity, specified that you could choose *20* numbers, one more than you really needed.

The official name for the technique that solves the socks puzzle is called the Pigeonhole Principle. The idea behind it is quite literal. If you want to put a bunch of pigeons into a bunch of boxes, and you have more pigeons than boxes, then at some point two or more pigeons must be placed in the same box. As principles of higher mathematics go, you can't get more straightforward than that. The concept has been traced to the *Schubfachprinzip* of Johann Peter Gustav Lejeune Dirichlet (1805–1859), a German mathematician whose career overlapped with that of Gauss.

But the simplicity of the Pigeonhole Principle—or, as it is sometimes called, the Dirichlet Box Principle—doesn't mean that it isn't potent. The applications of the Pigeonhole Principle are numerous, and the proofs usually involve figuring out what are the pigeons and what are the boxes. We won't give any more full proofs here, but the following are some challenges you might ponder if this subject has appeal:

> **Prove that there are at least two people in Tokyo with the same number of hairs on their heads.**

> **Prove that among any 16 integers between 1 and 30, at least two of them must differ by precisely 3.**

> **Prove that if you take at least 1 aspirin per day and a total of 45 aspirins over a 30-day period, there must be a stretch of consecutive days during which you take precisely 14 aspirins.**

As valuable as substitutions can be in the field of puzzle solving, I'm afraid that we found the technique through false pretenses. It turns out that "The Big Jump" wasn't about substitutions at all. Perhaps I should have made that admission sooner, but the delay was nothing compared to the one I experienced in real life. It took me at least 30 years after my initial exposure to "The Big Jump" before I realized the book's allegorical nature—that a series of baby steps could accomplish something enormous, as in a business career that started with a hot dog stand and culminated with the chairmanship of IBM. At this point it's probably too late for the hot dog stand, and the IBM position refuses to materialize, so I'll have to content myself with the thought that the serial jumping of little Ben can be put to work to solve puzzles that otherwise might be out of reach.

A few pages ago we encountered the Tower of Hanoi puzzle. We showed that we can go from the left picture to the right picture in just seven moves, where each move takes one and only one disc to a different post, at no time placing a disc on top of a smaller one.

There's actually no reason to restrict the Tower of Hanoi puzzle to just three discs. One of the legends surrounding the puzzle is that somewhere in an Indian temple, Brahman priests are busy moving *64* golden rings around three posts, and that when they finally complete the puzzle, the world will end. So, are we doomed?

At first glance it seems that we might survive on a technicality, because it looks impossible to handle so many discs without extra posts. But we can follow Ben's example and work our way from three discs to 64, one disc at a time. It's far easier than it sounds. If you had four discs instead of three, the puzzle could be solved by expanding the solution we already created, as follows: Use seven moves to place the three smaller discs onto the *middle* post. Now use a single move to place the biggest disk onto the right post, then seven more moves to place the three smaller discs on top of the largest one. That's 15 moves—2 × 7 + 1. The method can be repeated for five discs and so on, step by step (Figure 7-5).

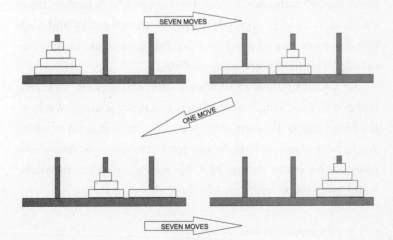

FIGURE 7-5

Using this recursive approach not only allows us to solve the puzzle for *any* number of discs but also yields a formula for how many moves the solution will require. The procedure we have just gone through illustrates that every time we add a disc, we double the number of moves, plus one. Note that for three discs, the number of moves, seven, can be expressed as $2^3 - 1$. For four discs, the number of moves is 15, which equals $2^4 - 1$, and the pattern continues. In particular, if you had 64 discs, the total number of moves would be $2^{64} - 1$.

The number $2^{64} - 1$ thus makes its second appearance in mathematical folklore. There's an ancient tale involving an inventor who seeks a fair price for something he created for the king. The king makes the mistake of allowing the inventor to name his price. Instead of asking for something obviously exorbitant, the inventor settles on the payment of one kernel of wheat on the first square of a chessboard, two kernels on the second square, and so on. The final square of the board will contain 2^{63} kernels of wheat, and if you add them all up—remember, we did create a formula for precisely this situation in Chapter 4—you get $2^{64} - 1$ kernels, far in excess of the world's supply. If the priests moved swiftly and made one move per second, their total solving time would be approximately 585 billion years, so no worries.

As a final example of recursion, suppose you were little Ben, living in a magic kingdom and hoping to prove yourself worthy of the king's puppy. The king gives you a challenge. Instead of asking you to jump atop his castle, he asks you to imagine a gigantic chessboard with a corner cut out of it. He asks you to fill in the chessboard using pieces that look like those shown in Figure 7-6.

FIGURE 7-6

The snag is that the chessboard he wants you to fill in has 128 units on a side. What do you do?

Fortunately, you remember that 128 is a power of two, so you think big but start small. If you had been asked to fill in a 4 × 4 chessboard that had a missing square, you'd have no problem, as shown in Figure 7-7.

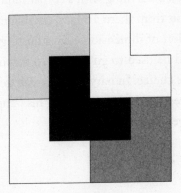

FIGURE 7-7

Now we build up, step by step. If we link four 4 × 4 tilings as illustrated in Figure 7-7, we have shown that we can tile an 8 × 8 square, as in Figure 7-8.

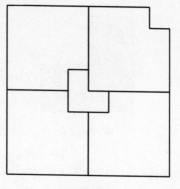

FIGURE 7-8

We can go from 8 × 8 to 16 × 16 in the exact same way, and so forth. Note that we didn't even need to bother constructing a 4 × 4 tiling in the first place. We could have started by noting that the L tromino is just a 2 × 2 tiling with a corner removed, and continued our induction from there. In fact, the diagram in Figure 7-8 makes no mention of dimensions. By adjusting the scale of the drawing, it could be used to go from 2 to 4, from 4 to 8, or any doubling of your choice. In particular, we can get to 128 in a few short steps, impressing the king, winning his puppy, and, no doubt, living happily ever after.

EIGHT

IT CAN'T BE DONE

A full century before Rubik's Cube became an international obsession, the 15 Puzzle owned the madness market. The puzzle consisted of 15 numbered wooden pieces sitting in a 4 × 4 grid, as shown in Figure 8-1. The unmarked region at the lower right represents an open space into which an adjacent piece can be moved, creating a new open space.

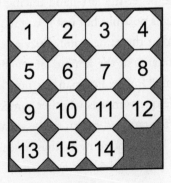

FIGURE 8-1

All is well in the figure except for the bottom row, where the 14 and 15 have been transposed. The challenge is to move the pieces around by sliding one square at a time until the 14 and 15 are in their proper places. The bad news is that the solution was not contained in the box. The even worse news is the reason why: There is no solution. The puzzle is impossible.

What a dirty trick, you may be thinking. Just as participants in the Boston Marathon insist on a finish line, the one thing we take for granted when we tackle a puzzle is that a solution exists. Why else would we voluntarily beat our brains out? The *Brooklyn Daily Eagle* was on this wavelength on the ides of March of 1880, near the height of the 15 craze: "It is understood that the punishment of felons in Sing Sing is made more horrible by furnishing prisoners with the sliding puzzle of 15." Did their snark betray a suspicion that the task could not be completed?

Although the 15 Puzzle is probably the most notorious impossible puzzle in history, there are others that are perhaps more familiar to recent generations of schoolchildren. Puzzles that involve tracing often fall into this category. A typical challenge would be to trace a diagram such as those in Figure 8-2 without taking your pencil off the page. The left-hand tracing is impossible. The right-hand tracing is possible, but not if you require that the tracing begin and end at the same point. Puzzles of this sort can be considered descendants of the famous Seven Bridges of Konigsberg problem, named for the Prussian city now known as Kaliningrad, in which the question was whether someone could traverse all the bridges in the city and return to the starting point without ever crossing the same bridge more than once. Fortunately, Konigsberg claimed Leonhard Euler as a native son, and through Euler's insights the bridge-crossing puzzle was shown to have no solution. Along the way he produced a complete classification as to which bridge layouts were negotiable and which were not.

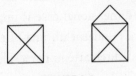

FIGURE 8-2

And now we see what we're up against. If impossible problems are presented to us as though they might be possible, and in real life it takes the likes of Leonhard Euler to demonstrate otherwise, it is no wonder that these puzzles can perplex one generation after another. To see some of the mathematics of impossibility in action, we'll take a painfully close look at another pencil-and-paper challenge that seems to reappear with every new generation: the water, gas, and electricity puzzle.

The objective of the puzzle is to draw paths connecting each of the three houses A, B, and C to each of the three utilities water, gas, and electricity without any of the paths crossing (Figure 8-3).

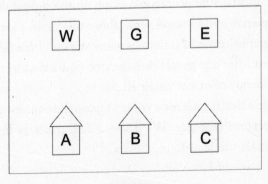

FIGURE 8-3

Save your pencils. You can't do it. That claim may seem surprising, because there's of course no problem in three dimensions: If two pipes cross, you just lift one of them up a little bit and you're all set. But the two-dimensional case is unattainable unless you route one of the pipes underneath one of the houses, as Henry Dudeney did in a last-ditch attempt at a solution.

FIGURE 8-4

The water, gas, and electricity puzzle is represented mathematically by the stick figure on the left of Figure 8-4. Denoted $K_{3,3}$, its technical name is the complete bipartite graph on three elements, meaning that every node in one set of three elements is connected to every node of another such set. Because this stipulation coincides with the requirements of the water, gas, and electricity puzzle, $K_{3,3}$ is also called the utility graph, and the impossibility of the puzzle is expressed in mathematical terms by the known fact that $K_{3,3}$ is nonplanar. The pentagonal graph on the right, called K_5, is also nonplanar, and in some sense those two graphs are it. Even though graphs come in infinitely many sizes and configurations, it turns out that any graph that can't be embedded in the plane contains a copy of one of either $K_{3,3}$ or K_5.

I confess that it took me a while to grasp the concept expressed in the previous sentence. What about the graph in Figure 8-5 (known in the trade as K_4)?

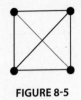

FIGURE 8-5

At first glance, there doesn't seem to be any way of drawing this graph in two dimensions without having the diagonal paths cross. But topologists are nothing if not sneaky, because they give themselves the right to bend objects to their advantage. Nobody said that you had to connect diagonally opposite points *within* the square, and if you take away that restriction you get the graph in Figure 8-6, where the intersection points are confined to the four vertices.

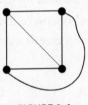

FIGURE 8-6

This same sort of topological trickery can be put to work to give the easiest known proof (albeit not a 100 percent rigorous one) that the water, gas, and electricity problem has no solution. Start by connecting each of the three houses to two of the three utilities. That situation can be modeled by a hexagon, as in Figure 8-7.

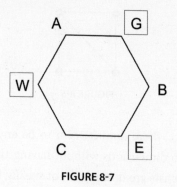

FIGURE 8-7

Obviously we can proceed a little further. We can connect house B to the water supply by drawing a line through the inside of the hexagon, and we can connect house C to the gas source by drawing a curved path on the outside of the hexagon, as shown in Figure 8-8. But there's no remaining path with which to connect house A to the electricity, because inside and outside are our only two options. Two's company, three's a crowd.

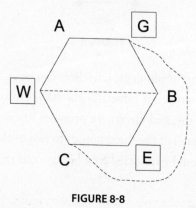

FIGURE 8-8

Yet sometimes an impossible challenge can be rendered possible by the simplest of adjustments.

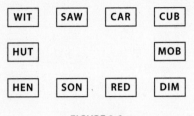

FIGURE 8-9

The rectangle in Figure 8-9 consists of 10 cards, each with a three-letter word printed on it. The task is to rearrange the cards so that each word card has precisely one letter in common with each of its two neighbors. A close look at the figure brings the unwanted news that SON and RED don't share a letter. In fact, no arrangement of the 10 cards will work, though some will bring you frustratingly close. Mathematically, a solution would be equivalent to a Hamiltonian Circuit (a path that travels to each vertex precisely once and returns to its starting point) in Figure 8-10, but none exists.

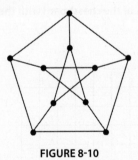

FIGURE 8-10

Fortunately, the graph can be easily modified so as to allow a Hamiltonian Circuit. In terms of the original puzzle, all you have to do to create an actual solution is to replace SUN by SON and HOT by HUT. The arrangement we're looking for is given in Figure 8-11.

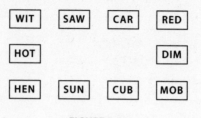

FIGURE 8-11

But puzzles that depend on sophisticated graphs with funny names by definition face a limited audience. The good news is that where impossible puzzles are concerned, even a small semantic twist can improve the entire dynamic, giving us a fighting chance. Consider the following classic (Figure 8-12):

You are given a chessboard from which two diagonally opposite corners have been removed. You also have 31 dominoes, each big enough to cover precisely two squares of the chessboard. Is it possible to cover the 62 remaining squares of the chessboard with the 31 dominoes?

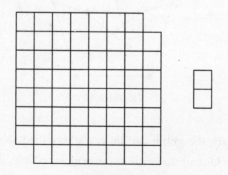

FIGURE 8-12

If you find yourself itching to start covering the big piece with dominoes, you've hopped on the wrong train. Try to focus instead on the puzzle's unmistakable kangaroo aspect, in the spirit of the internal clues we discussed in Chapter 2. Would the question be phrased this way if the covering were possible? Highly unlikely. And when you consider that this famous puzzle, dubbed the Mutilated Chessboard, has been used in corporate interviews, you know for sure that the answer to the question is no. Does the interviewer really expect the job applicant to verbally describe a 31-piece domino tiling on the spot? Of course not. Instead of searching in vain for a nonexistent tiling, we can turn our energy to proving that tiling the Mutilated Chessboard is impossible. There is a standard proof, and we'll arrive at it by borrowing a technique from Chapter 7 and looking at an analogous puzzle that was specifically crafted to help us out.

> In a small but very proper Russian village, there were 32 bachelors and 32 unmarried women. Through tireless efforts, the village matchmaker succeeded in arranging 32 highly satisfactory marriages. The village was proud and happy. Then one drunken Saturday night, two bachelors, in a test of strength, stuffed each other with pierogies and died. Can the matchmaker, through some quick arrangements, come up with 31 satisfactory marriages among the 62 survivors?

As long as gay marriage isn't an option—and at the time of the puzzle's construction, society had yet to explore in that direction—we can see at once that the matchmaker is doomed. The same logic applies to the Mutilated Chessboard problem, once we make the decisive recognition that (1) the two squares removed from the board are of the same color, and (2) the two squares covered by any individual domino must be of different colors. Try as you might, you'll never be able to cover the 62 remaining squares—32 white and 30 black—with the 31 dominoes (Figure 8-13).

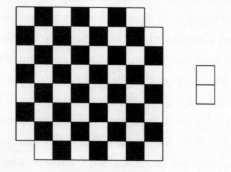

FIGURE 8-13

The matchmaker problem isn't a true isomorph because it isn't literally equivalent to the Mutilated Chessboard problem. For example, no one stipulated that the married couples had to be neighbors, the way they are on the board. Yet the matchmaker construct made the vital coloring insight impossible to miss, and reminds us that we would have been better off had our original chessboard been filled in with alternating black and white squares. A seminal 1990 experiment by Craig Kaplan and Herbert Simon of Carnegie Mellon University confirmed this advantage. Even better than having the board colored—black and pink, in their experiment—was to have the words "black" and "pink" written on the squares of the board. And best of all was to have the words "bread" and "butter" written on the squares, accentuating the natural pairing.

The key phrase in the original Mutilated Chessboard problem is "diagonally opposite." From a puzzle-setting perspective, those two words solve the problem of how to remove two squares of like color from the chessboard without really telling anyone. The same-color requirement is essential. Gomory's Theorem (1973) demonstrates that if you remove any two squares of *opposite* color from a chessboard, you can always tile the remaining 62 squares using 31 dominoes.

The mystery behind the infamously unsolvable 15 Puzzle can be unraveled in a similar stepping-stone fashion. This time our pathway puzzle will be one of my creations, a variant of a puzzle that I crafted for a miniature book called *Sit and Solve Brainteasers*. (The first edition of the book was shaped like a toilet bowl, to remind readers of the optimal solving locale.) In this puzzle, your objective is to transform the word RESIST into the word SISTER in precisely 12 moves, where a move consists of switching two adjacent letters (Figure 8-14).

RESIST → SISTER
FIGURE 8-14

You are, of course, welcome to carry out this assignment with Scrabble tiles. But before you dip into your game cupboard, note that the two chosen words aren't exactly arbitrary. They share the letters SIST in the same order, and a moment's reflection reveals the ease with which RESIST can be transformed into SISTRE. And another moment's reflection reveals that the obvious path to this transformation involves an *even* number of moves. For example, you can move the R from the far left to the far right in 5 moves, and then do the exact same thing with E, for a total of 10 moves. Unfortunately, SISTRE isn't a word, and it certainly isn't the target word. To get the target word, you have to transpose the

R and the E, creating an odd number of moves altogether. While I don't wish to open the mathematical Pandora's Box of Permutation Theory, suffice it to say that you're screwed. Any rearrangement of letters, numbers, symbols, or what-have-yous can be expressed as either an even number of swaps (called transpositions) or an odd number of swaps, but never both. In particular, if you can go from arrangement A to arrangement B in 11 moves, you can go back-and-forth a few extra times to make it 19 moves if you like, but a 12-move solution is unattainable.

But what if I said that the RESIST-SISTER Puzzle can in fact be solved? The key to the puzzle is that somewhere along the way you have to switch the two Ss. For example, you can start by switching the I and the S of RESIST, then switch the Ss, then switch the I and the S back again. You have reproduced the original word after an odd number of moves, at which point it is child's play to produce SISTER in an even (total) number of moves.

The 15 Puzzle offered no such cheap trick. To switch the 14 and 15 represents a single transposition—an odd permutation, if you will. But any sequence of moves that returns the open space to its starting point must be an even permutation. A rigorous mathematical proof would be somewhat longer than that, but that's the essence of the problem, beautifully masked by the puzzle's appearance and workings. (By contrast, the RESIST-SISTER Puzzle cannot avoid a reference to the odd/even issue, so the only question is which clumsy syntax to adopt.)

One of the amazing things about the unsolvability of the 15 Puzzle is that it didn't inspire cheating on a massive scale. Remember, unlike Rubik's Cube, in which the constituent "cubies" are all attached to the hidden hub in the center of the puzzle, the 15 Puzzle consisted of 15 free-floating pieces in a square shell. A solution therefore technically represented a sequence of moves to attain the desired position, rather than just the final position itself, which would have been easy to produce by simply picking up the 14 and 15 pieces and swapping them.

Perhaps the most inventive faux solution was made possible when the puzzle was produced with circular counters instead of square ones. Such a puzzle could be "solved" by a three-step process outlined in Jerry Slocum's magnificent book on the 15 Puzzle and repeated in Figure 8-15 with his permission.

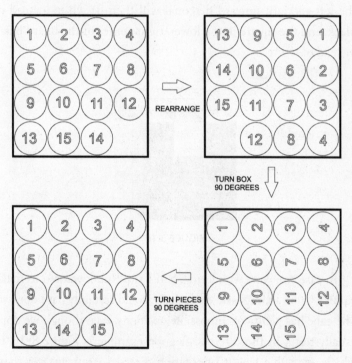

FIGURE 8-15

The key point is that the permutation that takes you from the original state on the left to the new position on the right *is* possible. From there it's just a matter of rotations. First you turn the box 90 degrees in a counterclockwise direction, producing the desired solution to the puzzle but with the pieces askew. Then all you have to do is turn the pieces clockwise by 90 degrees. The end result is a dead ringer for the actual solution unless the markings on the side of the box somehow betray the 90-degree rotation. (That's precisely the case in some trick chess puzzles, where the problem as stated is impossible unless you realize that the board has been turned 90 degrees. That tomfoolery is given away because the left-to-right upward diagonal will then be white instead of black, and the square in the lower-right corner of the board black instead of white.)

FIGURE 8-16

In case you were wondering, had Erno Rubik wanted to make Rubik's Cube an impossible puzzle, he could certainly have done so. The cube has plenty of impossible positions, even though we don't usually think in those terms. A good example is Figure 8-16, in which one cubie has its orientation reversed. You can't get there from the starting position unless you physically remove the cubie and replace it upside down. There are several different classes of impossible positions. Recall that if you took apart a 15 Puzzle and put the pieces back randomly, you'd have a 50-50 chance of being able to achieve the solution from there. If you took apart and reassembled a Rubik's Cube, your chance of being able to solve the puzzle is only 1 in 12. No doubt Rubik realized that his puzzle was plenty difficult on its own, and the world is grateful for the challenge.

NINE

SOLVED ACCORDING TO DOYLE

*How often have I said to you that when you have
eliminated the impossible, whatever remains, however
improbable, must be the truth?*

—SHERLOCK HOLMES, "THE SIGN OF THE FOUR"

Given that Sherlock Holmes apparently *never* said "Elementary, my dear Watson" until portrayed on the silver screen, the above quotation may be Sir Arthur Conan Doyle's most famous line. In and of itself, his advice is inarguable, and it has the added advantage of sounding smart without being pretentious. It's *reductio ad absurdum*, minus the Latin. The problem is that the advice is surprisingly difficult to put into practice when solving puzzles. But we intend to do just that, even if it takes us all chapter.

Perhaps the most frequent application of the Holmesian process of elimination is in trying to find some lost item around the house. The preferred strategy is to start with the most likely spot or spots and then move on to other areas if unsuccessful, but if you move on before absolutely, positively nailing down that the item in question was not in the drawer/shelf/cabinet/coat pocket/garbage compactor you looked at originally, you could be, um, a study in scarlet before the search is over. Some people are really good at scouring; some are not. If you have to retrace your steps five times before finding your quarry, you're in the latter camp.

The one thing we have to remember about Sherlock Holmes is that he was extraordinarily self-confident. If we aren't confident ourselves, our chances of mimicking him are slim to none, because the act of "eliminating the impossible" is itself impossible in the absence of conviction. The rub, of course, is that maintaining a high confidence level while tackling a fiendish puzzle is difficult by definition. The puzzle is *trying* to reduce you to a pulp.

Let's suppose you're a U.S. citizen who wanders onto Holmes's turf and attempts to solve the *Times of London*'s cryptic crossword. Trouble awaits. When unsuspecting Americans come across the clue "In banquets, no hot drinks will be brought round (4-3)," how are we supposed to know that the answer is "nosh-ups?" And whiffing on a Britishism isn't the worst of it. Once the threat of obscurity has contaminated two or three entries in a crossword puzzle, hapless solvers (such as yours truly) go into a tailspin. Beset by uncertainty and distrust, we find ourselves unable to figure out even the simplest clues. It's unacceptable to think that nosh-ups could take away our mojo, but that's what happens when we push ourselves to the brink. The real question is, how do we get it back?

In the late 1960s, researchers at the University of Pennsylvania conducted a set of experiments that produced a confounding result. In the initial segment, electric shocks were administered to two groups of dogs. One group could end the shocks by pressing a lever; the other group could not. Later on, the setup was altered, and each group was now given the chance to end the electric shocks by jumping over a low partition. Dogs in the first group had no problem doing so. But dogs in the second group, with few exceptions, didn't even try. Even though they held the power to eliminate their discomfort, they had already given up. One of the researchers, a graduate student named Martin Seligman, coined the term *learned helplessness* to describe this phenomenon of underperformance. If at first you don't succeed . . . the odds must be stacked against you, so why bother.

Several decades later, Charisse Nixon of the Pennsylvania State University at Erie conducted an in-class experiment, first suggested by psychologist David Myers, in which students were challenged to construct anagrams of three words, one at a time. Those who were able to form an anagram of the first word were asked to raise their hands. Roughly half the class did so. Then on to the second word, same rules. Again, half the class succeeded. Finally the class proceeded to the third word, for which the distribution of raised hands wasn't what you might have thought had you seen the words in advance.

Unbeknownst to the class, the students had been separated into two groups, with the left side of the room and the right side each getting its own set of words:

LEFT	RIGHT
Whirl	Bat
Slapstick	Lemon
Cinerama	Cinerama

Obviously the first two words given to the right side of the room had easy anagrams: "tab" for "bat" and "melon" for "lemon."

The third word was a bit tougher, but a little poking around reveals that "Cinerama" can be rearranged to spell "American."

The left side of the room wasn't so lucky. The words "whirl" and "slapstick" cannot be anagrammed, period, so not a single hand was raised from that side of the room during stages one and two of the experiment. Making matters worse, the left side had to endure the sight of a dozen or so raised hands from their class-mates on the *other* side of the room. The left-siders, by their own later admission, felt confused, rushed, frustrated, and, ultimately, stupid. So when it came time for them to complete the third ana-gram, they still couldn't compete with the right side, even though the challenges at that moment were identical. Professor Nixon had done to them what British crosswords do to me, though at least she had the courtesy to apologize.

Learned helplessness is a puzzle solver's worst enemy, no doubt, but it is also no stranger to the classroom. It is precisely the sort of collateral damage we've come to expect from a junior high or high school mathematics curriculum. Suppose you were asked to calculate the area of a triangle whose sides are 6 inches, 8 inches, and 14 inches. If you have painful memories lurking in your math-ematical past, you will bow out quickly. Whereas Renee Zellweger famously interrupted Tom Cruise in *Jerry Maguire* with "Shut up. . . . You had me at 'hello,'" with icky geometric calculations it's more like, "Shut up. . . . You lost me at 'triangle.'"

But doesn't a part of you wonder why you're being asked the question in the first place? There's a bit of Heisenberg Effect—or, if you prefer, kangaroo puzzle—brewing here, and we had best keep our wits about us so we can exploit it. Suppose I told you that you *can* solve the triangle puzzle.

And this one: What is the product of the digits of the number 61,285,390,334,596,127?

And even this one: What does $(x - a)(x - b)(x - c) \ldots (x - y)$ $(x - z)$ equal?

Hearing that a puzzle is solvable, or even hearing that someone else thinks we can solve it, helps make it solvable, period. If there is such a thing as learned helplessness, there must be something called assisted insight, and assists can be helpful even when the puzzle in question sits at the highest levels of mathematics. What set Fermat's Last Theorem apart from other adventures of number theory was the second part of Fermat's famous scribble: "It is impossible to separate a cube into two cubes, or a fourth power into two fourth powers, or in general, any power higher than the second, into two like powers. *I have discovered a truly marvelous proof of this, which this margin is too narrow to contain.*" By the time a century or two had passed, there was no one alive who actually believed the italicized claim, and the mathematics that eventually conquered the theorem would have been as foreign to Fermat as the space shuttle. But the sliver of hope created by Fermat's boast endured through the ages. Andrew Wiles, who conquered Fermat's Last Theorem a mere 358 years after the whole narrow-margins business, said, "I don't believe Fermat had a proof. I think he fooled himself into thinking he had a proof. But what has made this problem special for amateurs is that there's a tiny possibility that there does exist an elegant 17th-century proof."

The three problems I just posed aren't quite as challenging as Fermat's Last Theorem. I'm confident that you can follow the solutions to each of them—or maybe even solve them yourselves, especially when I hand you the final clue that the answers are all the same.

Because $6 + 8 = 14$, a triangle having those three sides is nothing more than two line segments lying on top of a third one. In mathematical parlance, such a construction is called a degenerate triangle, and its area is zero. And so is the product of the digits in 61,285,390,334,596,127, because one of those digits is a zero. The other digits don't matter at all. And even if the expression $(x - a)$ $(x - b)(x - c) \ldots (x - y)(x - z)$ conjures up unwanted visions of the quadratic formula, it's worth repressing those nightmares long enough to realize that one of the terms in that product is $(x - x)$, so the product equals zero for the same reason as the problem before. Trick questions can be the easiest of all—once you suspect that they are trick questions.

In practice, it's the mathematically gifted who fare better with these sorts of problems, despite the problems' eventual simplicity. That rich-get-richer dynamic is unfortunate, but there's no sense complaining about it, because we don't complain in other competitive arenas. If you're a golfer, you might lament the fact that the very best golfers get to play on the very best courses, thus magnifying their advantage, but you can hardly complain that they don't lift their heads up before they complete their putts, because you don't have to either. Professional tennis players are gifted, strong, and fit, qualities we'd all love to have, but should we view it as an unfair advantage that they also get to play with their knees bent, their eyes on the ball, and their racquets back?

If confidence building is our goal, it's worth a moment of our time to reveal that the rich don't necessarily get richer even when actual money is involved. Nobel Prize–winning economist Daniel Kahneman has cited a 2009 Gallup poll in which the happiness of respondents was graphed against their annual income. In one respect the graph was not surprising: Those with incomes below the $60,000 mark tended to be unhappy, and they got progressively less happy as their incomes declined. What was a surprise was that earning above the $60,000 mark did absolutely nothing to move the needle. The graph was flat. Money could alleviate misery, but it could not generate happiness—at least not within what Kahneman termed the *emotional self*. (Similar considerations eventually led Martin Seligman, the man who coined the term *learned helplessness*, to switch to a more positive platform. Having been drawn to psychotherapy as a means of identifying and treating depression, he came to the realization that although targeting the sources of depression can make a troubled patient less unhappy, you need different techniques to make an untroubled person more happy.)

Whatever the competitive arena, those at the very top got there by a combination of skill and technique. Skill is nice, but it is not readily transferable. Proper technique, on the other hand, is available to anyone, even though sometimes we have to watch the very best to remind us of what we could be doing better. With puzzles, the ability to analyze a problem is aided by a comfort with its trappings. All we really need to achieve our maximum is a capacity to stop, think, even breathe our way through. As long as we don't panic, we're way ahead of the game. And we're also in a position to make full use of Holmesian logic.

Consider the following practice puzzle:

The members of the 1970s band Hamilton, Joe Frank & Reynolds ("Don't Pull Your Love," 1972; "Fallin' in Love," 1975) go out to dinner, and the total bill, including tip, is $99. How can they pay for the dinner using only a combination of $20, $10, $5, and $1 bills, assuming that each of the band members has to pay the same amount?

It seems clear that each of the band members should pay $24.75, so there's a bit of a problem here. If you want to conclude that the puzzle is impossible, you'll probably have some company. But Holmes would disapprove. Once you conclude that the puzzle is possible, then it cannot be the case that the band has four members. And it doesn't. Joe Frank was one person, namely Joe Frank Carollo. With only three people in the band, it's a simple matter to divide the $99 check by having each of them pay $33. Obviously the puzzle is much more difficult when communicated orally, because you have no punctuation to guide you. And as long as you pronounce the group's name in a normal fashion, it's difficult to conclude that only three people are involved. With that warm-up in hand, I invite you to identify the impossible and the improbable in the following three puzzles from my private stash.

The first is a geometric dissection puzzle I dreamed up for the 10th Gathering for Gardner in March 2012. The Gathering has been held in Atlanta every other year or so, courtesy of such puzzle people as Elwyn Berlekamp, Tom Rodgers, and Mark Setteducati. The honoree Martin Gardner was very much alive when the celebrations began as a supposedly one-time event in 1993, but in later years Gardner didn't attend, and the 2012 event was the first since his passing in 2010.

My puzzle, dubbed Penta-gone, consisted of 12 plastic pieces nested in a white plastic frame with an inner rectangle, as shown in Figure 9-1.

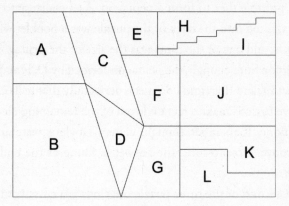

FIGURE 9-1

The objective of the puzzle was to rearrange the pieces so that they formed a rectangle and a pentagon—simultaneously, of course.

The interlocking of pieces H and I may be familiar from our discussion on dissections from Chapter 2. It seems fair to divulge that indeed part of the solution involves dropping the I piece down a notch to form an even thinner rectangle. One wrinkle that might not have been apparent to the recipients of my puzzle, even though geometric puzzles are in their wheelhouse, is that this part of the dissection doesn't quite work. Remember when I said in Chapter 2 that the stairstep or zigzag method depends on the rectangles being in the proper proportion? Well, these rectangles aren't quite in the proper proportion here. But they're very close, so close that the distance separating them from the perfect proportion is less than the margin of error you need to fit them into the frame in the first place. That the pieces jostle around ever so slightly is expected and necessary. The naked eye simply cannot tell that the fit isn't perfect.

But the naked eye *can* reasonably tell that pieces D, F, and G can be put together to form a pentagon. As a courtesy to the recipients, fitted into the back of the puzzle was a booklet with several hints, and one of the hints was the size of the pentagon I was looking for. Sure enough, the pentagon formed by D, F, and G was a perfect fit, and this time I do mean perfect. By that reckoning, all you have to do is make a rectangle out of the remaining pieces and you're done. Remember, though, Conan Doyle is watching your every move. No pressure. The answer is found at the end of the chapter.

The second of the three puzzle offerings is a crossword puzzle I constructed several years ago. It features a British-style pattern, meaning that the intersections between across and down entries occur with less frequency than is the case with American crosswords. Life is thereby easier on the constructor but tougher on the solver. But I promise that obscure entries such as "nosh-ups" are nowhere to be found. Here goes (Figure 9-2):

ACROSS

1. Anagram of "spacer"
4. Plays the guitar
8. Dark liquid found in a well
9. Legendary Egyptian queen
11. The ultimate, in a way
12. A real pickle
13. ___ King Cole
15. Classic hair-removal brand
16. Leap ___
21. Curve
23. ___ Friday
24. Multi-time NCAA basketball champion

26. AKC-registered breed [two words (5,4)]

27. Anagram of "ate"

28. One of the four seasons

29. Long-time Yankees stalwart who debuted in 1995

DOWN

1. He performed with Nash

2. Earns, as in a lot of money [two words (5,2)]

3. ___ of the litter

5. This became the bestselling album of all time by an individual performer

6. ___ Station (Washington, DC, train spot)

7. These are positioned at the bases of Christmas trees

10. Olympic fencing subspecialty

14. A world capital

17. Type of cheese

18. One of Asia's longest rivers

19. Loosely worn cloth garment

20. Lipton competitor

22. Wood finish option

25. Darth Vader adversary

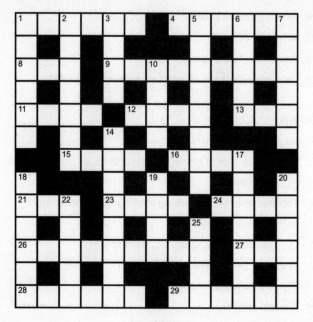

FIGURE 9-2

Exactly where one should start a crossword puzzle is a matter of personal choice. It would be nice to always be able to fill in 1A, but that's not realistic, and it's certainly not possible here: The word *spacer* has many possible anagrams. So we just go to the low-hanging fruit and work from there. For this puzzle it feels reasonable to go after 9A, then use a few key letters to get 3D and 5D, then work from there to get 1A and 1D. Figure 9-3 shows what the puzzle would look like if we took all those steps:

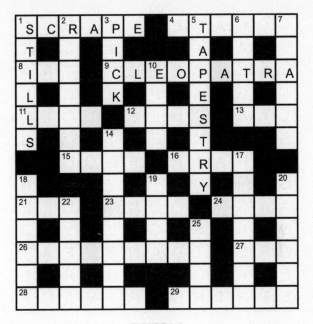

FIGURE 9-3

If the solving process we just started is taken to its logical conclusion, however, we run into a big problem. We'd love to have the word STRUMS at 4A, but we can't very well push the M over one space and replace it with a U (Figure 9-4). Can you overcome this problem and solve the puzzle? Again, Holmes is watching, and Watson probably is, too. Like the dissection puzzle that came before it, the answer to this puzzle can be found at the end of the chapter.

FIGURE 9-4

We now crank things up a notch. Our final Holmesian puzzle grew out of something I came across as I was doing research for *Number Freak*, a book I wrote in 2008. The nature of the research was to compile interesting properties of the whole numbers from 1 to 300. There was, of course, much more to say about the smaller numbers than the larger ones, so the fact that there are 292 ways of making change for a dollar came in awfully handy toward the end. But a funny thing happened long before then, when I was assembling factoids for the number 36.

Obviously the number 36 has all sorts of arithmetic quirks about it. Being 6 × 6, it is a perfect square. It is also a triangular number, in that if bowling pins were arranged in eight rows instead of four, you'd need 36 of them to fill the triangle. The Lincoln Memorial has 36 columns, one for every state of the union at the time of Lincoln's death. And so forth. But I was chagrined to come across a conundrum called the 36 Officer Problem. Posed by the great Leonhard Euler in 1782, it asked the following:

> **How can a delegation of six regiments, each of which sends a colonel, a lieutenant-colonel, a major, a captain, a lieutenant, and a sub-lieutenant, be arranged in a regular 6 × 6 array such that no row or column duplicates a rank *or* a regiment?**

The dressed-up version of the story, not entirely supported by the historical record, is that Euler had traveled to St. Petersburg and was asked this question in person by Catherine the Great. Whatever the problem's underpinnings, by now we're sufficiently trained to ask, "What's with the choice of 6 × 6?" Is there something special about that particular size?

Clearly there's no problem with, say, 3 × 3, as Figure 9-5 shows. If the three ranks are A, B, and C and the three regimens are plain, dark gray, and light gray, each rank belongs to a different regimen in all three rows and columns. The letters form what is known as a Latin square, even though there's nothing Latin about it—the name derives from Euler's habit of using Latin letters when constructing squares with that property. Latin squares have proven useful in experimental design and even athletic competitions, and the concept was put on the map once and for all by Sudoku and KenKen puzzles, which by definition form Latin squares but upon their completion. In Figure 9-5, not only do the letters form a Latin square but so do the three backgrounds, and every combination of letters and backgrounds is represented. This type of composite construction is known as either a Graeco-Latin square or an Euler square.

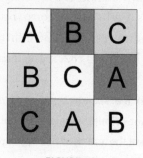

FIGURE 9-5

Oddly, though, there *is* a problem with the 2 × 2 case, as can be seen in Figure 9-6. When we fill in the left-hand column in generic fashion, as in the left-most drawing in the figure, we have no options for the second column: If we put A on the top and B on the bottom, the two rows each repeat the same letter, and if we put B on the top instead, then both Bs are light gray and both As are plain. It doesn't work.

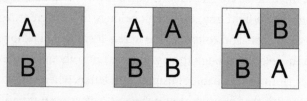

FIGURE 9-6

The answer for the 6 × 6 case wasn't furnished until 1901, when an amateur French mathematician named Gaston Tarry demonstrated that the desired arrangement is not possible. And some 60 years after that came this amazing discovery from Ernest Parker, Raj Chandra Bose, and Sharadchandra Shrikhande: Other than the trivial 2 × 2 case, the 6 × 6 case is the *only* one that doesn't work. That's right. You can find Euler squares measuring a billion units on a side, but 6 × 6 is out of the question.

■

The reason for my chagrin was that here I was in the 21st century, 100 years after Gaston Tarry's lifetime achievement, and I had never heard of him or the problem he solved. Maybe that's because the 36 Officer Problem had been put to rest so many years before—unlike, say, the Four Color Map Theorem, a longstanding conjecture that was solved in my lifetime. But as I emerged from my self-induced shame I couldn't help but think that the 36 Officer Problem could somehow be converted into a puzzle.

I had been down that road before. Some years earlier I had made a dissection puzzle out of the marvelous equation $3^3 + 4^3 + 5^3 = 6^3$. Using the admittedly suboptimal building blocks of colored dice and Krazy Glue, I fashioned a 15-piece puzzle in which the 3 black pieces formed a cube of side 3, the 4 blue pieces formed a cube of side 4, the 8 red pieces formed a cube of side 5, and all 15 pieces together formed a cube of side 6.

I would soon learn that there are more efficient ways of accomplishing this specific dissection. Eight-piece dissections exist and are known to be minimal (because cubes have eight corners and each of the corners in the large cube must come from a different puzzle piece). You need nine pieces if you insist that each piece be shaped like a block, but I was hopeful that the funny-looking pieces in my dissection would at least give it some character. I took my puzzle to rural Newton, New Hampshire, the unlikely site of the U.S. headquarters of German game and puzzle company FX Schmid (now Ravensburger/FX Schmid). Once inside I met a young man named Mark Hauser, who in rejecting the puzzle gave a terrific piece of advice that has followed me around ever since. "What customers want," Hauser said, "is a puzzle that looks easy but turns out to be hard." He pivoted, pointed to my puzzle, and said, "That *looks* hard." Of course he had me dead to rights. I had thought that by creating a laddered sequence of puzzles, I had addressed the difficulty issue, but the final set of pieces still had a daunting look. I showed the puzzle to a couple of companies thereafter but eventually retired it to my office game shelf, the island of unwanted prototypes.

With the 36 Officers Problem, I invite you to sit back and follow my logic, for better or worse, for a puzzle that *did* make it to market. From the beginning, the puzzle I envisioned consisted of a base that resembled a city skyline, consisting of 36 towers of six different heights. In our newfound lingo, the tower heights formed a Latin square. My father, then 83 years young, took no time at all in dubbing the model "Latin Manhattan" (Figure 9-7).

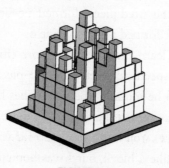

FIGURE 9-7

On the base were to be placed 36 pieces of six different colors, each color having six different heights. The objective of the puzzle was to place the pieces so that all 36 pieces attained the same height, and each row and column contained each of the six colors: red, yellow, blue, orange, green, and purple. If you think of the colors as representing the six different ranks and the base towers as the six different regiments, the 36 Officer Problem was thereby modeled in three dimensions. And if all I had wanted to do was to create an impossible puzzle, I could have stopped right there. Gaston Terry had done the groundwork for me.

What I had in mind, however, was a puzzle that was more than an all-or-nothing proposition. Even if you couldn't place all 36 pieces on the base without duplicating a color along a row or column, surely you could do 20, 24, or some in-between number. It turns out that the number of pieces you can legally place varies considerably depending on the structure of the Latin square that you choose as a base. For some bases, you might not be able to go higher than, say, 26. But others, including the one in the diagram, take you all the way to 34.

With the help of a couple of engineers and a 3D printer, I was able to get a model made in time for Toy Fair 2007, where I had an appointment with the one company that figured to be the best match: ThinkFun of Alexandria, Virginia. ThinkFun had been the leader in mathematically pedigreed puzzles ever since the husband-and-wife team of Bill Ritchie and Andrea Barthello founded the company (as Binary Arts) in 1985. One of their earliest efforts, a creation of the memorably named inventor Ferdinand Lammertink, was Topspin, a permutation puzzle that evoked the 15 Puzzle of yesteryear, except that it could be solved. Covering all bases, ThinkFun eventually purchased the rights to the 15 Puzzle, this time in a multichallenge form in which all the desired states of the puzzle could actually be achieved. You get the point. I had come to the right place.

For puzzle inventors, Toy Fair is a Dickensian opportunity, being the best of times to find the top people in the industry and the worst of times to sustain their attention. I had a half hour session booked with ThinkFun, and even those 30 minutes were shaved at the edges, but finally Bill Ritchie and I sat down together in their clear-walled conference room, one step removed from the skiff-skaff bustle of the Javits Center.

The meeting went well. I had brought two other creations along for the ride, a complex marble maze and a tetrahedral puzzle that refused to stand up straight, but those sacrificial lambs had been tossed aside long before my 30 minutes were up. If there are 50 ways to leave your lover, there are twice as many ways for a game or puzzle to be rejected, and the lines are as familiar to me as "The Star Spangled Banner." *It's already been done. It's not fun. It's too big. It can't be mass-produced. It doesn't fit with our line.* So with every new meeting I know that my audience, whether it's one person or three, is juggling all these alibis at once. But Ritchie liked what he saw, and despite all the distractions of the moment and the rejection checklist that was surely swirling in his head, he had noticed his staff and other onlookers peering in to our little fish bowl to see what the model was all about. It was colorful and intriguingly shaped, and it didn't even look hard. He asked me to send it down to Virginia for further review. And somehow I got out of there without ever revealing the solution.

In July 2007 I took a trip to the ThinkFun offices and explained how everything worked. I talked a bit about Euler and the 36 officers and all that, and I started putting colored towers onto the base according to the rules of the puzzle. With just a couple of pieces to go, my task appeared hopeless; but I dropped the last one in, and Bill Ritchie's eyes popped out as if on stalks. You see, the puzzle *was* solvable. Despite all appearances, it wasn't actually isomorphic to the 36 Officer Problem. That vital hurdle having been cleared, the project got its long-awaited go-ahead. Many fits and starts later, the puzzle was launched in late 2008 as 36 Cube.

The gimmick that made the puzzle work was simplicity itself. I mentioned earlier that the base I chose enabled a solver to place 34 pieces without doubling up on any row or column. But what does a 34-piece partial solution look like? From a puzzle constructor's perspective, they're very easy to create. All you have to do is insert a 2×2 "knot" into the puzzle—just imagine Figure 9-5 with the backgrounds representing towers of different heights and the letters representing pieces of two different colors, each with the appropriate height—and then arrange things such that every other piece can be placed legally around the knot. And that's what I did. There are myriad ways in which an attempted solution can go sour, but the 2×2 case is sort of the canonical knot, much like the "utility graph" $K_{3,3}$ in the world of impossible two-dimensional graphs. The final flourish was to make sure that there were plenty of 2×2 knots in this particular base, guaranteeing that near misses would be a commonplace event. The puzzle can be solved by (1) identifying the pattern of the near misses (where 34 pieces can be placed but no more), and then (2) identifying the one 2×2 knot that can be "untied," an impossibility in the pencil-and-paper formulation of Latin squares but quite possible within the three-dimensional representation. I realize that this explanation is incomplete, but the vow of secrecy I took back in 2007 has yet to be lifted. As Euler could have said to Catherine the Great but probably didn't, tough noogies.

The final step is to view the puzzle from a marketing perspective. The public doesn't know about the 36 Officer Problem, so they won't know that the puzzle is theoretically impossible. Do you tell them that the puzzle is in fact solvable? My opinion was that you shouldn't, for reasons that we've already encountered on several occasions: The information that a puzzle is solvable helps make it solvable in real time. The powers-that-be at ThinkFun overruled me on that one, presumably because it's *their* inbox, not mine, that gets stuffed with customer hate mail. And, really, what are they supposed to do when customers call? They can't very well lie, unless they have a peculiar wish to make 36 Cube their final product. So they had a choice between giving thousands of angry callers the solvability message over the phone, one by one, or notifying everyone at once on the puzzle box. In that sense they certainly made the right choice. Our compromise was to break from all established post-1880 tradition and not include the solution with the puzzle itself.

I suppose part of me longed for a mania along the lines of the 15 Puzzle, but that mania depended on an ill-informed public. It's not that people were mathematically ignorant back in 1880—far from it. The presidential election of that year was won by Congressman James A. Garfield, who four years earlier had published a trapezoid-based proof of the Pythagorean Theorem, so there was brainpower to spare at the highest levels. The pace of information flow, however, was another matter. Professors William Woolsey Johnson and William Story addressed the impossibility of the 15 Puzzle in the December 1879 issue of *American Mathematical Monthly*, but for some reason the publication of that particular issue was delayed by several months. It wasn't received by the Library of Congress until April 1880, and in the meantime the 15 Puzzle was flying off the shelves. And there were just enough rumors of actual solutions to keep the craze going.

Times have changed, and those luxuries don't apply in the 21st century. The 36 Cube has spent its entire life knowing that at any given moment, some spoilsport could post the solution on the Internet. In fact, such postings have appeared with some regularity, though credit should be given to the many true sportsmen who went to great lengths to camouflage the solution while still telling the world of their triumph. For some reason the puzzle did especially well in Germany, making the Top Ten Spielzeug list for 2009. A friend pointed out that by making a bigger splash in Germany than in the United States, I had something in common with David Hasselhoff. To which I can only say that as long as we're not talking about solving the 15 Puzzle, anything is possible.

SOLUTIONS TO EARLIER PUZZLES

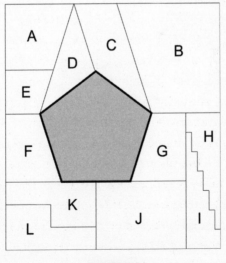

FIGURE 9-8

FIGURE 9-9

TEN

WHEN INDUCTION GOES BAD

I intend to live forever. So far it's working.

—STEVEN WRIGHT

How wrong is Wright? The truth is that misguided prognostication plagues us as soon as we are hatched. An infant's cries may be triggered by physical pain or discomfort, but they are prolonged by a lack of awareness that the pain will not last forever. When that awareness finally kicks in, however many years of real inductive experience might be required, everyone involved is happier, and the concept of "forever" now takes on more allure.

Wright's hope for eternal life rests on the slender thread that the basis for future projections is his own unblemished track record rather than the miserable track record of his ancestors. Surely he's aware that even people with names built for longevity, like Pliny the Younger, came up short. But with stakes this high, it's worth exploring what puzzles have to say about the whole business of taking a pattern and extrapolating into the future. As always, we start easy.

> **Mary's father has five daughters. Four of them are named Nana, Nene, Nini, and Nono. What is the name of the fifth daughter?**

Okay, maybe that was too easy. How about this?

> **What is the next number in the following sequence? 1, 2, 4, 8, 16, ?**

Unless you simply don't trust the person who's asking, the answer is clear. Each number in the sequence is twice its predecessor, so the question mark should equal 2 × 16, or 32. But there is an alternative and by now classic interpretation of this sequence that yields a very different result. Take a circle and draw two points somewhere on its circumference. By connecting those two points with a line segment, you have divided the circle into two parts. If you add another point on the circumference and connect all possible pairs of points with line segments, you can create four regions within the circle. Adding another point makes it possible to create eight regions, and a fifth point makes it 16, as in Figure 10-1.

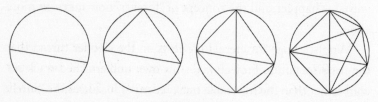

FIGURE 10-1

It's tempting to believe that this pattern goes on forever, doubling the number of regions with each additional point, as if we were cutting pieces of cake and stacking them on one another at every turn. So it's no small surprise to hear that the sequence strays from our ideals when we go from five points to six. Try as we might, the maximum number of regions generated by six points is 31, as in Figure 10-2, one short of the 32 regions we might have expected. So close and yet so far. From there the pattern breaks down completely, as the next four terms of the sequence are 57, 99, 163, and 256. The matching of the first five members of the sequence was merely a coincidence, not a blueprint for extrapolation. The Strong Law of Small Numbers of famed number theorist and mathematical gamesman Richard Guy assures us that this sort of thing happens all the time, because "There aren't enough small numbers to meet the many demands we make of them."

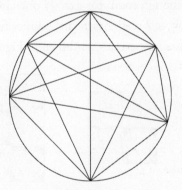

FIGURE 10-2

There's actually an even simpler way of devising a sequence that begins with 1, 2, 4, 8, 16, 31. All you have to do is list the proper divisors of 496, which form the set {1, 2, 4, 8, 16, 31, 62, 124, 248}. Is it cheating to use a finite set? Maybe, but this one is interesting mathematically because the sum of its elements equals 496. A number that is the sum of its proper divisors is called a "perfect number," and 496 is the third such number, after 6 (= 1 + 2 + 3) and 28 (= 1 + 2 + 4 + 7 + 14). Believe it or not, no one knows whether the total number of perfect numbers is finite or infinite. At this writing, 47 perfect numbers have been discovered, but they will not be listed here: The largest of the lot has 25,956,377 digits.

What about this sequence? 1, 4, 9, 16, 25, 36, 49, ?

The seven listed numbers look very much like the first seven perfect squares, so we expect to see the question mark occupied by $8^2 = 64$. But we could get a different result by defining the nth member of this sequence as the number of circles that be packed into an $n \times n$ chessboard, where the diameter of a circle equals the side of one of the squares on the board. Say what? Well, a 1×1 chessboard has enough room for exactly one circle; that much is obvious. Similarly, a 2×2 chessboard admits four circles, and a 3×3 board admits nine circles, as shown in Figure 10-3.

 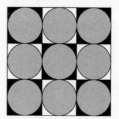

FIGURE 10-3

This exercise seems silly. What possible difference could there be between the number of circles that fit on the board and the number of squares on the board itself? The answer is that the circles don't have to go inside the individual squares. When we get to $n = 8$, as in a normal chessboard, all of a sudden we have room, even if barely, to pack the circles in a different way (Figure 10-4).

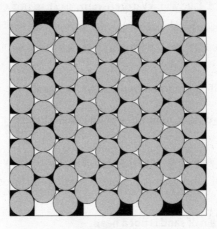

FIGURE 10-4

If you count up all the circles, you'll get five columns of eight and four columns of seven, for a total of 68 circles, not 64. From there the sequence diverges slowly but steadily, as in 1, 4, 9, 16, 25, 36, 49, 68, 86, 106, and so forth. William of Ockham would giddily point out that you have to go through an awful lot of trouble for the original sequence not to be what it looked like in the first place. But puzzles thrive on the exceptions, and if you are getting suspicious about the whole business of induction and extrapolation, your misery is only beginning.

Eubulides of Miletus, a contemporary of Aristotle, is credited with a conundrum now known as the Sorites Paradox. Its peculiar name (pronounced so-RIGHT-eez) derives from the Greek word for "heap." The original formulation of the paradox began with an enormous heap of sand, from which grains were removed one at a time. When at long last only one grain remained, what you had was no longer a heap. Of this there can be no doubt, but when exactly did it cease to be a heap? The matter has been adjudicated by the likes of Ludwig Wittgenstein, so it is not considered trifling in philosophical circles.

Let us present the paradox in different terms.

1. **One billion grains of sand constitute a heap.**

2. **A heap of sand minus one grain is still a heap.**

Both propositions seem airtight, but jointly they suggest that a single grain of sand is a heap. Let's try going in the other direction:

1. **One grain of sand is not a heap.**

2. **Adding one grain of sand to something that is not a heap cannot make a heap.**

This time we're stuck with the conclusion that even a trillion grains of sand is not a heap. This same framework has been applied to the number of hairs on a man's head, the continuum of shades of red, and even to the number of seconds until old age or outright expiration. Whatever the underlying medium, there comes a point where things change. The concept of a straw that could break the camel's back may have seemed odd when we first encountered it, but the other side of the argument is even worse.

In mathematics, the logic of an infinite progression of this type is formalized by something called an inductive proof. As we foreshadowed with the recursive constructions in Chapter 7, such a proof takes on the following form: First, prove that something is true for $n = 1$. Then prove that its truth for n implies its truth for $n + 1$. If you can do both of these things, the statement is true for all n. This description isn't known to win students over, so induction is sometimes presented in the form of an infinite ladder.

If you are capable of reaching the first rung of that ladder, and if you can demonstrate that you can always advance to the next rung, no matter which rung you happen to be on right now, then you can climb the entire infinite ladder.

The technique of induction is typically invoked in proving formulas such as the fact that the sum of the first n positive integers equals $n(n + 1)/2$. This formula renders explicit the insight, introduced in Chapter 7, of young Carl Friedrich Gauss. He was given the problem for $n = 100$, and he quickly observed that the first 100 positive integers consist of 50 pairs that each add up to 101, and that's exactly what the formula says. Mathematical induction can take us to far greater heights, starting with the formula that the sum of the first n squares equals $n(n + 1)(n^2 + 1)/6$. Or maybe even the sensational result that the sum of the first n cubes equals $(n(n + 1)/2)^2$, which we can now identify as just the sum of the first n integers squared.

On a lighter note, mathematician George Pólya used faux induction in an effort to prove that there was no such thing as a horse of a different color. His argument that all horses must have the same color went something like this: If you have just one horse, clearly that horse is the same color as itself. Now assume that any group of n horses must be the same color. By the basic principle of induction, we're done if we can show that the same must be true for $n + 1$ horses. Let's create two smaller groups from those $n + 1$ horses. The first group consists of horses 1 through n, while the second group contains the horses 2 through $n + 1$. There are n horses in each group, so the induction assumption indicates that the horses in each of the two smaller groups must be of the same color. But these two subsets clearly overlap, so all $n + 1$ horses must be the same color, completing the proof.

Pólya's tongue-in-cheek effort falls flat among those who have never seen a mathematical proof using induction, because the whole thing looks so weird. But the proof's flaw is fairly subtle given how unsubtle the error of the conclusion is. The folly that dooms Pólya's proof is that the step from $n = 1$ to $n = 2$ fails, because in that case the two sets {1} and {2} do not overlap. All Pólya's argument really says is that if any two horses were the same color, all horses would be the same color. The logic is impeccable, but it's just not so.

As difficult as induction is when presented rigorously, it comes to life when embedded within a puzzle.

> Two mathematicians, whom we'll call X and Y, are on a train. Each of them thinks of a positive integer and whispers it to a fellow rider, Z. After a short while Z gets up and announces, "This is my stop. You have each thought of a different number, and neither of you can deduce whose number is larger." Mathematician X, whose number was 62, thinks along the following lines: "Obviously Y didn't choose 1. If he had, he'd know that my number was larger, given Z's statement that we chose different numbers. But Z said that we couldn't figure out which number was larger. Just as obviously, Y knows that I didn't choose 1. Our choices must therefore have come from the subset {2, 3, . . .}. But by the same reasoning as before, that would mean that neither of us chose 2. By mathematical induction, all positive integers are eliminated—which is completely absurd!"

This induction can't possibly be correct, but figuring out exactly where it went off the rails (as it were) isn't so easy. We start by asking whether the initial statement is true. In other words, can we actually rule out the number 1 as a choice for either X or Y? The answer is yes. Obviously it originally might have been the case that X or Y, in an unimaginative mood, chose the number 1, but Z's statement makes it clear that neither one did so. As we've seen, had either mathematician chosen the number 1, they would have had an automatic inference that the other person's number was higher.

Having ruled out 1, can we proceed with the induction? In other words, is 2 the new 1? Again, yes. If either X or Y had chosen 2, the fact that 1 has been ruled out does in fact put them in a position to infer that his number must be the smaller of the two, and again that contradicts what Z said. Surely something happens with 3 that will stop this stupid induction dead in its tracks. Not really. Had either man chosen 3, they'd know within a very short period of time that the other number must be larger, because the only alternative is that the other guy chose 2, in which case he would have spoken up right away. This is getting worrisome. Inductions just don't peter out, but if this one doesn't peter out awfully quickly then we're in big trouble, because we know that two numbers *were* chosen.

The reason the induction implodes is that it involves a hidden clock. At each stage, the set of deductions required to eliminate a given number grows, perhaps only in milliseconds if the mathematicians are sufficiently keen, but it grows nonetheless. Ruling out 1 takes no time at all, but ruling out 2 takes an extra mini-beat as the practitioners follow the logic we just outlined. By the time you reached 62, or whatever, you'd have long since lost track of the microunits that defined each step of the induction, so you'd be in no position to judge what number was larger. And if you did happen to know the precise temporal flow, yes, you'd be able to figure out whether your number was larger, because all you'd be doing is counting, but then the paradox is gone. Besides, Z said you couldn't do that, so there.

The idea that perfect logic can have unexpected consequences is hardly new. Perhaps the oldest and most famous example is the donkey placed precisely in the middle of two mounds of hay. With no logical reason to prefer one hay mound to the other, the donkey starves to death. This construction dates back to Aristotle but the formal name of the paradox is Buridan's Ass, named sarcastically for the 14th-century French philosopher and moral determinist Jean Buridan.

Buridan's Ass actually has applications in modern times. The artificial intelligence effort to produce chess-playing computers is best-known for Deep Blue's head-to-head matches against Garry Kasparov in 1996 and 1997, but a generation earlier the bugs were still being worked out. In a match between COKO III and Genie in the second U.S. Computer Chess Championship in 1971, COKO as white assumed a commanding position, and black, apparently not having been programmed to resign hopeless causes, threw his remaining pieces in front of the white onslaught to produce the improbable position shown in Figure 10-5, with white to move.

FIGURE 10-5

Obviously white can checkmate by moving his queen to b2 or his bishop to c4. But it seems that COKO was ill-equipped to deal with such an embarrassment of riches. Lacking a specific instruction to administer one of the two death blows, white did neither. COKO ended up moving its king up and down between c1 and c2, seemingly toying with its prey but actually stuck in a cretinous loop. As COKO fiddled and its programmers burned, Genie's king-side pawns advanced, promoted themselves into queens, and won the game.

The ghost of Buridan's Ass also haunts so-called arbiter circuits of asynchronous computing systems. Arbiters are responsible for making decisions when two or more processors request access to shared memory, and delays can arise if these requests are too close together. Just as two people pause when they simultaneously converge on a doorway, the arbiter may need to pause before it can attain a stable state from which to make a decision, *because a decision must be made.* In 2003, a design flaw along these lines (called a "race condition") in a Unix-based GE energy management system delayed the activation of an alarm that might have headed off that year's great North American blackout. Fortunately, that particular bug has been addressed, and in general progress in circuit design has been such that Buridan-type delays are now measured in femtoseconds, a trillion of which amount to a mere thousandth of a second.

Here's a classic induction puzzle built from the oddities of perfect logic:

There is an island upon which a tribe of 1,000 people resides. The tribe's religion forbids them to know their own eye color, or even to discuss the topic; thus, each resident can see the eye color of all other residents but has no way of discovering his or her own (there are no reflective surfaces). If a tribesperson does discover his own eye color, the island's religion compels him to leave the island at noon the following day. All the tribespeople are highly logical and devout, and they all know that others are also highly logical and devout (and they all know that they all know that each other is highly logical and devout, and so forth).

Of the 1,000 islanders, it turns out that 100 of them have blue eyes and 900 of them have brown eyes, although the islanders are not initially aware of these precise statistics (each of them can of course see only 999 of the 1,000 tribespeople).

One day, a blue-eyed foreigner visits the island and wins the complete trust of the tribe. One evening, he addresses the entire tribe to thank them for their hospitality. However, not knowing the local customs, the foreigner makes the mistake of

mentioning eye color in his address, remarking "how unusual it is to see another blue-eyed person like myself in this region of the world."

What effect, if anything, does this faux pas have on the tribe?

The easy answer is "nothing." Followed by, "How could the innocent comment of a grateful visitor have any effect on the tribe when the comment didn't tell them anything they didn't already know?"

That's a reasonable point of view, but suppose that only one member of the tribe had blue eyes. In that case the visitor's comments would have been news indeed to the individual with blue eyes, who would have left the island the following day at noon. What if two people on the island had blue eyes? In that case neither of the two would leave the next day, because they didn't have an automatic inference of their own eye color. But when noon struck and neither one of them had left the island, each could make the inference that the other one wasn't the only person with blue eyes. Because everyone else on the island was known not to have blue eyes, that left only them. The result? Both would leave on the second day. And so on. The answer to the puzzle, surprising as it may seem, is that on the 100th day, all 100 people with blue eyes would leave the island.

Note that this puzzle, by forcing the action to take place at a specific time of day, takes care of the temporal ambiguity of its predecessor. What isn't resolved is the question of how on earth the visitor's innocent comment led to anything. The answer relates to something called "common knowledge." We know what common knowledge means in real life: something that everyone knows, such as the fact that if you're standing on the South Pole, you're in Antarctica. Mathematically, however, for a group of people to have common knowledge of a fact requires that when they all know it, they all know that they all know it, they all know that they all know that they all know it, and so on, ad infinitum.

This is sort of a strange concept. The closest approximation in real life would be the final two players at the World Series of Poker, where Player A tries to gauge the hand of Player B, who in turn makes inferences based on the strategy of Player A, who in turn knows the likely range of B's responses, and so on, at least a few rungs into the infinite ladder. The possibility of a bluff throws a monkey wrench into the logical chain, but the island puzzle is complicated enough as it is. The bottom line is that until the visitor's fateful expression of thanks, the fact that there were blue-eyed people on the island was known by all, but this knowledge didn't satisfy the rigorous definition of common knowledge.

■

Of course, when it comes to false induction puzzles, the granddaddy of them all is the Unexpected Hanging Paradox:

> On Sunday evening a judge tells a condemned prisoner that he will be awakened and hanged on the morning of one of the following five days. The judge says that it will happen unexpectedly—that is, the prison will not know that the hanging will occur until the moment the attendants arrive. But the prisoner soon convinces himself that no such hanging is possible. After all, if the judge sets Friday as the morning of the hanging, the prisoner will know it on Thursday night because he is still alive and will realize that the next day is the final day of

the execution period. Such foreknowledge is contrary to the judge's specifications, so Friday is ruled out. But if Friday is ruled out, then, by process of elimination, so are Thursday, Wednesday, Tuesday, and Monday. Of course, on Wednesday, the prisoner is hauled out of bed, much to his surprise, and hanged. What went wrong?

This puzzle first appeared in the early 1950s and has undergone more critical examination than any other conundrum of that era, unless you count *The Catcher in the Rye.* In 1994, by which time the puzzle had chalked up no fewer than 57 references in the academic literature, Dean Clark of the University of Rhode Island, writing in *Mathematics Magazine*, summed up 40 years of exasperation: "It is unclear which domain of human intelligence should take custody of it. Is it a problem in pure logic? Semantics? Psychology? Probability theory? Is it a problem without a solution or with multiple solutions?"

For what it's worth—recognizing that my feeble contribution is unlikely to alter the academic balance on the subject—I was always wary of the prisoner's induction on the grounds that it doesn't kick in until the very last day, at which point you have to travel backward in time to rule out the preceding days. But I don't pretend the problem is quite that simple. In certain circles, the logical angle is entirely subservient to the semantic angle. The thought there is that the judge's claim about the hanging being a surprise amounts to a self-referential paradox. In some analyses this angle gets carried out to a logical near-absurdity, wherein the seeming contradiction of the prisoner not expecting the hanging is resolved by concluding that the prisoner has *already* rightly inferred that the hanging is impossible, and therefore it *does* come as a surprise when it actually occurs, even on the last day. Endowing a condemned prisoner with this much mental acuity seems odd, but it is just as Samuel Johnson predicted: "Depend upon it, sir, when a man knows he is to be hanged in a fortnight, it concentrates his mind wonderfully."

When it comes to judges and sentencing, a source of more recent academic inquiry is the fact that timing can be everything. A team led by Shai Danziger of Ben Gurion University tracked the results of over 1,000 parole-board hearings in Israel, measured over a 10-month period. The judges typically heard between 14 and 35 cases each day. Each case averaged about six minutes, and the judges took two meal breaks during the day. It turns out that the likelihood of parole starts at about 65 percent each day, then steadily declines until the morning snack. After the judge returns, the likelihood of parole vaults back to the 65 percent mark, then declines until lunch. After lunch it's the same story: The likelihood of parole returns to its elevated starting point, then falls rapidly until hovering just above zero. Figure 10-6 lays it all out. While it's tempting to thank your lucky stars that you're not in an Israeli prison, the effect that the graph describes is surely universal. At what time of day is your job interview? When exactly does your child's college application get decided? How depressing to think that the turning points in our lives can be linked to someone else's blood-sugar level.

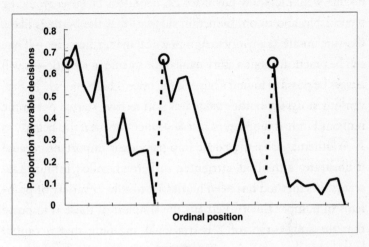

FIGURE 10-6

Robert Louis Stevenson's wonderful short story "The Bottle Imp" revolves around an induction paradox. The setup is that the main characters are presented with the opportunity to buy, for a price negotiated with the previous owner, a bottle containing a genie. (In Stevenson's construct, the bottle's previous owners included such luminaries as Napoleon and Captain Cook.) Once you purchase the bottle, the genie will, of course, fulfill your every desire and make you successful and happy and all that, but there is a catch. The catch is that you must resell the bottle for a lower price than what you paid. You needn't bother selling it for a higher price, because it will magically return to you. And if you still have the bottle when you die, you are cursed to Hades forever, which as we've already established is a very long time.

Given these conditions, a familiar sequence presents itself. Obviously no one would buy the bottle for 1¢, because it would then be impossible to get rid of. But if no one would buy it for 1¢, then no one would logically buy it for 2¢, because a 1¢ buyer would be unavailable, and so on. Induction suggests that the bottle is like a Corvair, unsafe at any price. As a practical matter, however—if one can be practical about a story named for a genie in a bottle—it will always be possible to find a buyer if the price is high enough. In Stevenson's story, the bottle's value was said to have started out in the millions but had dropped to just $80 when the narrative began.

Mathematically, the bottle imp paradox is important because it illustrates a method, attributed to Fermat, called Infinite Descent. This method has been harnessed to solve a variety of problems in number theory. Perhaps the simplest of these is to prove that the square root of 2 is irrational, meaning that it cannot be expressed as a fraction. This fact was apparently discovered by the Pythagoreans—to their evident horror, because the concept of an irrational number was foreign and an affront to their sensibilities. The way the proof works is that if you found numbers p and q such that p/q equaled the square root of 2, you could use that equation to find another solution using numbers *smaller* than p and q. Well, you can't keep on doing that forever, so the initial discovery must have been impossible. A similar logic can be applied to show that the equation $x^4 + y^4 = z^4$ has no solutions in positive integers.

Well, this is a trifle complex, you may be thinking. But, just to make a point, we're going to use Fermat's Infinite Descent to prove the following theorem together:

Theorem: Take any positive integer and count the number of letters in its (English) name. Take the resulting number and count the number of letters in its name. Keep going. No matter what number you started with, you'll end with . . . 4.

Reread if you like, because stranger-sounding theorems are tough to come by. But the proof is remarkably simple. We start by noting that the number of letters in "four" equals 4, so if you happened to have chosen 4 originally, you'd be done in one step. Also note that if you chose 3, the sequence would next take you to 4 in two steps, first to 5, because "three" has five letters, and then to 4, because "five" has four letters. And if you had chosen 1 or 2, well, each of those has three letters, so you'll arrive at 4 having taken an additional step.

This is no way to prove anything, is it? Yes, we've covered the numbers 1, 2, 3, and 4 (and actually 5 along the way) without much trouble, but we've got an infinite road in front of us. Except that the infinite road can be handled in just one extra step. The elementary coup de grace is that for any number greater than or equal to 5, the number of letters in that number is less than the number itself. (Usually *way* less. A number such as three million six hundred twenty-four thousand nine hundred fifty-one is a mouthful, but it requires only 59 letters, and 59 is a whole lot less than 3,624,951.) So no matter how big a number we select, the process of counting the letters brings it down, and down, and down, and, having already taken care of the very lowest cases, eventually to 4. We are done.

There is one teensy-weensy wrinkle worth pursuing. Suppose that Spanish, not English, is your preferred language. Suddenly "four" isn't so special anymore because the Spanish word for "four," *cuatro*, has six letters. But there is a fixed point in Spanish, namely *cinco*, which has five letters and means "five." Is the theorem we just proved true upon replacing *English* with *Spanish* and *four* with *cinco*?

Not quite. Yes, a whole lot of numbers will converge to *cinco*, and the principle that the number of letters in a number is less than the number itself is every bit as true in Spanish as it is in English, once you get beyond *cuatro*. But suppose your original number was 15, or *quince*. The number of letters in "*quince*" equals 6, or "*seis*." The number of letters in "*seis*" equals 4, or "*cuatro*." The number of letters in "*cuatro*" equals "*seis*." Oops. We've entered a black hole. So the Spanish version of the theorem must undergo a minor but vital alteration. Either the sequence ends at *cinco* or it bounces forever between *cuatro* and *seis*!

As "The Bottle Imp" unfolded, Stevenson built a couple of loopholes into his original paradox. Toward the very end of the story, the hero, Keawe, purchases the bottle for a penny, willingly dooming himself to display his love for the story's heroine, Kokua. But Keawe was later able to give the bottle to a sailor who figured that he might as well take it because hell was on his horizon no matter what he did for the rest of his life. Before that resolution, Keawe and Kokua had tried to sell the bottle in a foreign currency, where the division of a penny into centimes could effectively reset the paradox in someone else's hands. Their efforts had the flavor of a reverse stock split, but we'll close this chapter by showing that you don't need to resort to such esoterica to apply the bottle imp to the stock market.

Day-in, day-out, the market gives investors a dilemma of bottle imp proportions by placing a positive value on enterprises whose long-term existence is highly questionable. These sorts of situations pit sentimental, wishful thinking long-term investors against aggressive, opportunistic short-sellers—investors who sell stocks on margin, hoping to buy them back at a lower price—and you can probably guess the likely winner of that matchup. Suppose you had held shares of Blockbuster Video in 2005. The good news was that

Blockbuster had the highest market share in its business. The bad news was that the future of the bricks-and-mortar video rental business was very much in doubt. Netflix had been around for five years. Video streaming of movies was not yet widely available but was considered inevitable. These events didn't go unnoticed by Wall Street. Blockbuster shares traded at about $10 per share in early 2005, down from $30 a few years before. But the question remained whether that discounted price was a good one.

If you use the same inductive, hot-potato argument that applied to the bottle imp, the answer would be no. The nontrivial likelihood that Blockbuster would be worth close to zero in the future was a pretty strong argument not to own it in the near-term at any price. But the whole point of the bottle imp paradox was that the imp would have been accorded some value despite the inevitable death spiral of selling at a loss. The stock market analogue is that at any given moment, even a doomed company can provide plenty of reasons to own it—or, if you prefer, plenty of distractions from the case *not* to own it. Maybe the company's business, however grim its future prospects, is a cash cow. Maybe its cash flow results in a healthy dividend, which hasn't been cut *yet*. Maybe investors just think they'd be terribly unlucky not to have some greater fool offer a higher price than what they paid.

A discounted price is a powerful allure, even if it applies only to the rearview mirror. As we now know, the only profitable action an investor could have taken with Blockbuster in 2005 was to have sold it short. But to consider a short sale you'd have to get over the fact that the stock had already declined so steeply. This is a generic hurdle of the short-selling trade, and it can be overcome by invoking one more puzzle:

You buy 100 pounds of potatoes and are told that they are 99 percent water. After leaving them outside, you discover that they are now 98 percent water. How much do they weigh now?

This puzzle is known as the Potato Paradox. It merits the paradox designation because the answer, 50 pounds, seems way low. But the arithmetic is quite straightforward. Originally the potatoes consisted of 99 pounds of water and 1 pound of whatever else makes up a potato. If that 1 pound now represents 2 percent of the total weight, that total must be 50 pounds.

Translated to the stock market, if you buy a stock for $10 and sell it for 50 cents, your percentage loss is 95 percent. Had you instead waited until the stock had reached $5 before making your purchase (you canny investor, you), your percentage loss would have been only . . . 90 percent. You still lost your shirt. The more optimistic viewpoint is that if short selling is in your bag of tricks, you still made a killing even if you missed the initial downdraft.

By the way, because it's poor form for a book chapter to leave unfinished business, the answer to the puzzle on page 160 is "Mary."

ELEVEN

YOU ARE HERE
The Search for a Fixed Point

The German Gestalt psychologist Karl Duncker (1903–1940) spent a mere 37 years on this planet before deciding that he had had enough, but he left a supply of conundrums and posers that is sure to torment several generations of fine minds. A glimpse into his thinking style is afforded by an early effort called the Radiation Problem, which gained a significant following in cognitive psychology circles and which goes like this: Given a patient with an inoperable stomach tumor, and lasers that destroy organic tissue when operating at sufficient intensity, how can a hospital cure the patient with these lasers without harming the healthy tissue that surrounds the tumor?

Anyone who devised laser problems before 1940 must have been ahead of his time, and the Radiation Problem has proved to be a difficult test. When presented under laboratory conditions in the absence of any helpful hints, success rates have been as low as

10 percent. The solution is to fire low-intensity lasers from several different starting points so that the rays converge at the tumor. The idea is that no individual laser beam will destroy healthy tissue but the combination of lasers will prove deadly at their point of convergence. Divide and conquer. Duncker would be pleased to know that this type of thinking is no longer fanciful or theoretical. Many cancers are approached with a combination of treatments rather than a single miracle drug. The concept of an HIV cocktail was borne in the 1980s and continues in the form of so-called highly active antiretroviral therapy, with different drugs applicable to different stages of the HIV life cycle. And it is not unusual for children recovering from certain types of surgical procedures to be given alternating doses of sedatives and muscle relaxants. Both classes of drugs accomplish the objective of keeping the patient asleep but in very different ways, and an alternating regimen avoids the problems inherent in a sustained dosage of either form of postoperative pain management.

In this chapter we will focus on two of Duncker's other trademark challenges. The second of these will be familiar to puzzle buffs, but the first one is a less widely known puzzle that, like the Radiation Problem, dates back to Duncker's days as a graduate student at Clark University. The task he concocted was to attach a candle to a wall using only the objects shown in Figure 11-1.

FIGURE 11-1

You could try lighting the candle and creating some wax, as some of Duncker's subjects did, but none of them enjoyed enduring success via that route. The solution isn't particularly difficult, but it requires converting the box of tacks into a shelf, as shown in Figure 11-2.

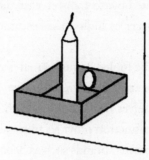

FIGURE 11-2

If you found the solution immediately, congratulations are in order, because in Duncker's experiments you wouldn't have had much company. What makes the problem harder than it looks is "functional fixedness," a term Duncker coined to express the difficulty human beings have in altering the known function of an object, even when the problem before them requires such a step. Where lateral thinking fails, functional fixedness thrives.

Conjurers and illusionists are surely grateful that their audiences come with this built-in deficit. When a magician lights a candle and simultaneously removes a coin hidden in the box of matches, or takes a drink of water and discretely moistens his fingertips, the ruses are successful because the functionally fixated audience doesn't question the real purpose of the props. True, any audience member who watches five successive performances might wonder why the Great Santini always gets thirsty at a specific moment of the illusion, but absent such curiosity the behavior is seldom questioned. The Central Intelligence Agency seemed to understand this phenomenon from the get-go. Shortly after its formation in 1953, the CIA recruited famed British-born magician John Mulholland, long-time editor of the journal *The Sphinx*, to write a manual that would teach men and women of international intrigue how to transfer nasty liquids into cups and to perform other pieces of high-stakes escamotage while in plain view of their marks.

If the KGB can be fooled by functional fixedness, there is little hope for the rest of us. The truth is that we humans need fixed points to guide ourselves through life, and the idea that a specific object has a specific function provides tremendous simplicity. It is no accident that consumer brands such as Scotch tape and Kleenex have been able to attain generic status. The entire population, minus a few antitrust lawyers here and there, prefers it that way.

The ability to resist functional fixedness usually takes one of three specific forms. The first is brilliance. Thomas Edison, predictably, was immune. Legend has it that visitors to Edison's winter residence in Fort Myers, Florida, complained for years about the sluggishness of his front gate, only to find out that the gate was connected to a pump that filled Edison's private swimming pool. No functional fixedness there. The second is necessity. NASA engineers were able to save *Apollo 13* by finding alternative uses for the duct tape and other items on board the spacecraft, literally squaring the circle (or vice versa) in refitting the straight-edged carbon dioxide filters into the round receptacles of the craft's lunar module. The third is youth. According to a clever experiment devised by psychologists Tamsin German and Greta Defeyter, we all have a chance at avoiding functional fixedness—as long as we do it before age six.

In German and Defeyter's experiment, young children were asked to help a puppet reach a high shelf by creating a tower from foam blocks stored in a box. The required insight was to use the box itself as part of the tower, and in finding that solution five-year-olds outshone seven-year-olds, the latter group being distracted by the box's original function. (When the box was presented *separate* from the blocks, the solution came swiftly for all age groups, just as Duncker's candle problem is rendered far easier when the tacks are introduced outside the box rather than within it.)

As we grow older, we discover that the concept of a faux fixed point takes on a completely different meaning, because life introduces us to the concept of generational misconceptions. For example, to those born in 1954, as was this author, it seemed reasonable to conclude circa 1960 that the purpose of television was to enable people to watch westerns: *Gunsmoke*; *Bonanza*; *Have Gun, Will Travel*; *Rawhide*; and *Tales of Wells Fargo*, to name a few. *Cheyenne*, *Maverick*, and *The Rifleman*, to name a few more. And that's not even including *The Lone Ranger* or *Death Valley Days*, and we still haven't listed so much as half of the 26 westerns that dotted the prime-time landscape at the peak of the genre. But six-year-olds weren't positioned to understand the "peak of the genre" stuff. Whatever was, was, and was presumed to be eternal. Yes, there was something ironic about the idea of television providing a fixed point in 1960, because TV itself was the new kid on the entertainment block. Come to think of it, the idea of anything being fixed on the basis of a human being's observations is silly, given that the human being making the observation is a transient phenomenon. But we of course prefer to think of ourselves as fixed.

Anyone who had been familiar with entertainment cycles over the years could have alerted me that westerns were simply having their day in the sun. And as long as fads of the 1950s were being laid bare, perhaps someone could have clued me in that tail fins weren't an essential part of an automobile's aerodynamics. No doubt you could add your own examples of things you mistakenly thought of as permanent based on childhood observations. I know I'm not alone: Screenwriter Nora Ephron once assumed that the ability to solve acrostic puzzles was a guaranteed right of adulthood just because her mother was a whiz at them.

Life's fixed points come in many different forms, and they aren't all as contrived or dopey as the ones we come up with individually. The investment world, for instance, is loaded with them, and the ability to distinguish what is temporary from what is enduring is the name of the game. A few generations ago, sage investors would have told you that dividend yields on stocks should be greater than bond yields. There was even a solid reason available—stocks were riskier than bonds, and investors needed some compensation for adopting those risks, which sounds like a timeless guideline along the lines of "hit the ball with the bat." However, investors who came of age during the 1970s and 1980s saw a very different picture. That generation would have felt that bond yields had to be *higher* than stock yields because bonds didn't offer the same potential for capital appreciation. Bonds had in fact yielded more than stocks since 1958, and the idea that this was a fixed condition gained considerable traction during the Internet-stock boom, when the average dividend yield of the S&P 500 dipped to unheard-of lows, hovering barely over 1.0 percent in the year 2000. Then, in 2008, to surprisingly little acclaim, dividends inched up, bond yields inched down, and the situation reversed itself for the first time in

50 years. The late Peter Bernstein, long-time professor, investment manager, and author of the acclaimed 1996 risk treatise *Against the Gods*, lived through both ends of that 50-year gap, and his words on the subject are tailor-made for the issue at hand: "A profound philosophical dilemma is present here. The yield inversion in the spring of 1958 taught me a lesson I have never forgotten: anything can happen. Just because a relationship had held since the beginning of time is no reason to believe it would also hold until the end of time."

Not all fixed points are creatures of our imaginations. To find where real ones come from we must entertain a mathematical diversion, courtesy of one final puzzle from Karl Duncker's personal collection. This puzzle has the feel of something that has been around since medieval times, and readers who have seen it before may be surprised to learn of its 20th-century roots, but here is the puzzle in its abridged form:

> A monk walks up a mountain trail beginning at 9:00 a.m. He reaches a temple at the summit at 5:00 p.m., at which point he settles in for the night. The next day he walks down the mountain beginning at 9:00 a.m., reaching his original starting point at 5:00 p.m.
>
> Is there a time at which the monk is in precisely the same spot on both days?

Think about it. Doesn't your gut tell you that the answer must be yes? That's the direction in which most people lean, even if they can't quite articulate why. The usual kangaroo element applies: If the answer is no, why is the question being asked in the first place?

But the one thing we know for sure is that we can't compute

the answer. If we are to solve the puzzle, we must do so in the form of what mathematicians would call a nonconstructive or existence proof. Yes, if the monk had the decency to travel at a constant rate, he'd be halfway up the mountain at 1:00 p.m. on both days, and we'd have found our fixed point. In real life, though, ascending is a tad more difficult than descending, and the unabridged problem makes it clear that no such shortcuts are allowed:

> **The monk ascended the path at varying rates of speed, stopping many times along the way to rest and eat the dried fruit he carried with him. He reached the temple shortly before sunset. After several days of fasting and meditation he began his journey back along the same path, starting at sunrise and again walking at variable speeds with many pauses along the way. His average speed descending was, of course, greater than his average climbing speed.**

At this point a little doubt sets in, just as Karl Duncker would have predicted or even insisted on. What if the monk raced up the mountain for the first three-quarters of his climb, paused for an hour of prayer, then remained on his hands and knees as he crawled his way to the summit? Might he somehow stagger his speeds in this fashion to avoid the fixed point?

The answer is no. A fixed point—that is, a point the monk occupies at precisely the same time of day, coming and going—must always exist, and the reason for this is stupefyingly simple. Instead of thinking about the monk making his return trip on some day subsequent to his ascent, imagine a second monk descending the mountain on the very same day. A moment's thought reveals that the original puzzle is equivalent to asking, in this new setting, whether the two monks will ever cross paths, and the answer is of course a resounding yes.

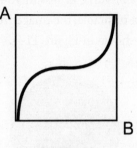

FIGURE 11-3

Figure 11-3 translates the monk problem into a simple graph. The upward-sloping line represents how the monk's progress up the mountain might have looked had he gotten off to a quick start, slowed down in the middle, then enjoyed a burst of energy at the end, though his precise speeds don't matter a whit. The problem essentially asks whether it's possible to go from point A in the upper left to point B in the lower right without crossing the upward path. In this formulation, not only is the answer clearly no but the problem isn't very interesting and is a mild insult to our intelligence. Mathematically, all that it is required is that the monk moves *continuously* up and down the mountain, and that's guaranteed by the conditions of the problem. The monk cannot step off the path, and he cannot dissolve into nanoparticles that instantaneously reassemble themselves somewhere in the distance. The bottom line is that his path to the summit, however it is shaped, can be traced without removing our pencil from the paper, and as long as that's the case for the two lines, they will intersect.

If you thought we were done with fixed points, you're sadly mistaken. Not only that but you're in for a miserable several pages, because we're going to look at fixed points from every conceivable angle, both mathematical and personal. We'll start by turning back the clock to the days of vinyl records. If you placed an LP record onto the spindle of a turntable, the image from above would look something like that shown in Figure 11-4.

FIGURE 11-4

If you give the record a spin, every single atom of the record will be in a different place from wherever it started unless you somehow spin the record precisely 360 degrees or a multiple thereof. But if the record didn't have a hole in it, the above statement would no longer be true, because the very center of the record would never move (assuming that you figured out how to spin the record without a spindle).

The three-dimensional version of this principle is even more significant mathematically. You start with a cup of coffee and you stir it. You can keep stirring as long as you want, long after the sugar and cream or whatever you might have added has been thoroughly mixed, and guess what? There is guaranteed to be some point in the coffee that is in its original place. We won't be able to identify this point the way we could with the center of the vinyl record, but it is there. According to legend, a mathematician named Luitzen Brouwer made this very conjecture about stirred coffee sometime in the early 20th century. His observation gained mathematical legitimacy in 1912, when he published a paper that would become known as the Brouwer Fixed Point Theorem, one of the most important theorems in the field of topology.

Nonmathematicians can have difficulty absorbing the language of Brouwer's theorem. Technically, to guarantee a fixed point you need a continuous function acting on a closed, bounded, convex set, and it's not surprising that this language is deemed unfriendly by some. Or many. Or most. The fact that spinning a record or stirring a cup of coffee satisfies those conditions is of little solace. Gruesome jargon affects other, similar results that nonmathematicians could appreciate if only they had a translator on hand. When mathematicians say, "There does not exist an everywhere nonzero tangent vector field on the 2-sphere," they mean that at any given moment there exists a point on the surface of the earth with zero horizontal wind velocity. This result is better known at the local level as guaranteeing an eye in the hurricane, the existence of which is fodder for *The Worst-Case Scenario Survival Handbook*: Knowing that even the most severe windstorms contain a placid patch makes it less likely to confuse the arrival of the eye with the end of the storm. In this case the underlying mathematics at least tries to be friendly. The result about vector fields is known in the trade as the Hairy Ball Theorem, because it is mathematically equivalent to saying that you can't comb a head of hair without creating a whorl somewhere.

The truth is that many real-life representations of fixed-point concepts are surprisingly easy to comprehend. We have all seen a "You Are Here" sign at an airport terminal, for example, but few would recognize that the X in the sign is an example of a fixed point. All the mathematical conditions are satisfied: Every point in the airport is associated with a point in the map, the mapping is surely continuous, it is surely closed and bounded, and, best of all, points of interest are even closer together in the sign than they are in real life—that's what mathematicians call a contraction mapping. Therefore, a fixed point exists. Well, of course it exists. The X in the sign is the same thing as the X in real life.

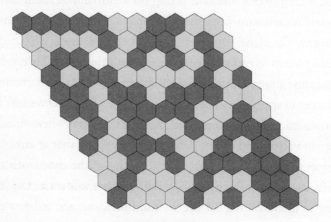

FIGURE 11-5

Shifting gears for just a second, the game of Hex is played on a parallelogram-shaped grid filled with hexagons. Players take turns placing hexagons of two different colors on the spaces in the grid. The objective is to form a complete path of one color. In the game shown in Figure 11-5, black is trying to connect the north and south sides of the grid, while gray is trying to create an unbroken east–west path. It may not surprise you to hear that Hex can never end in a draw. In the game in Figure 11-5, for example, whichever color is placed in the open white square would create a win for that color. The fact that the game can't end in a draw was first demonstrated by John Nash of *A Beautiful Mind* fame and, in this context, is mathematically equivalent to the Brouwer Fixed Point Theorem. A strange-looking result, until you consider the resemblance between the black and gray paths and the line up the mountain in Figure 11-3.

Fixed-point theorems of a more advanced sort have a variety of applications in economics and game theory. These theorems can guarantee, under certain conditions, that there is an equilibrium between supply and demand or between the optimal strategies of different players playing a complex game. The famous Nash Equilibrium of game theory was built on a 1941 paper by Shizuo Kakutani, which in turn extended Brouwer's Fixed Point Theorem to so-called set-valued functions. Three decades later, your author had the privilege of taking several high-octane undergraduate mathematics courses from Professor Kakutani, a diminutive man whose energy in front of the blackboard not only astounded students one-third his age but also left him covered in chalk dust at the end of each lecture. During one memorable post-class conversation with Kakutani, he mentioned to me that John von Neumann was the smartest person he had ever met. I laughed inwardly, knowing that the smartest person *I* had ever met was standing there right in front of me. Call it a different form of "You are here."

The message, then, is that fixed points are abundant under tightly defined geometric conditions. But the fixed points of geometry can face a tough time when they meet up with the functional fixedness of the human mind. To confront this lofty topic, we'll take a look at the first important fixed-point exploration conducted by mankind—the search for the center of the universe.

There were never more than two viable candidates for the universe's center: the earth and the sun. The geocentric theory held that the earth was at the center, and it is no surprise that this theory came first. Our planet certainly feels as though it were standing still, not hurtling through space at 66,000 miles per hour. But the heliocentric theory, which places the sun at the center, is older than you might think, having been introduced by Aristarchus of Samos circa 300 BC. This theory gained little adherence in its day. Not only was it counterintuitive but it was specifically counter to the prevailing mythology, in which Helios commandeered a chariot that took the sun around the earth each day, foreshadowing the difficulties that heliocentrism would face in what became an 1,800-year quest for legitimacy.

From a technical standpoint, the heliocentric theory seemed flawed because of a phenomenon called stellar parallax: If the earth rotated about the sun, why didn't the position of the stars change from year to year? But the geocentric models of the universe were suspect for their inability to account for other things that eluded the naked eye, notably the phases of Venus and the retrograde motion of the planets. Occam's Razor devotees will see the larger pattern of a simple theory breaking down under repeated stress. Eventually—and here *eventually* is clearly the operative word—heliocentrism gained favor. In 1543, a dying Nicolaus Copernicus published his *De revolutionibus orbium coelestium* (On the Revolutions of the Celestial Spheres), which placed the sun at the center, albeit with the other planets making circular orbits instead of the elliptical orbits revealed by more modern science. By the early 1600s, when Galileo's telescopic observations of the moons of Jupiter pushed him firmly into the heliocentric camp, the momentum had shifted for good.

However, by this time, if not before, the alternative fixed point to the heliocentric theory wasn't the geocentric theory; it was the Church. The Catholic establishment of that era wasn't ready for scientific advances that departed from biblical passages such as Psalms 93:1 ("He has fixed the earth firm, immovable") or Psalms 104:5 ("Thou didst fix the earth on its foundation so that it never can be shaken"). Scripture was a different sort of fixed point from the one that had prompted scientific inquiry, but in the short term it could be far more powerful.

The long-term conflict between heliocentrism and scripture had all the makings of the immovable object versus the irresistible force, the storied battle between two fixed points that can't exist physically and can't coexist logically. This battle in turn has its roots in the Omnipotence Paradox, which asked if God could make a stone so heavy that he could not lift it. And we are reminded that the Chinese word for *paradox* is literally "spear-shield," dating back to a bit of ancient folklore in which a man tried to simultaneously sell a spear that could pierce any shield and a shield that could resist any spear, until asked what would happen when the two came in contact. The modern version of this stalemate, one in keeping with the Duncker puzzle that started this entire discussion, would be the unresolved standoff between Dr. Seuss's North-Going Zax and his South-Going Zax. The point is that some battles don't have winners.

It would be tempting to say that heliocentrism won out. Its biggest boost came in the mid-1800s, by which time telescopes had become powerful enough to detect and measure parallax, which, as widely suspected, had gone unproved for a millennium or two, not because the stars didn't move but because they were simply too far away for primitive instruments to measure. But because heliocentrism's competition was a human rather than a geometric fixed point, its victory was and remains incomplete. Even in the 21st century, according to a 1999 Gallup poll and confirmed by subsequent studies, one in five Americans believes that the sun revolves around the earth. In Russia, that figure is one-third.

■

Dueling fixed points appear in contexts far more modest than the structure of the solar system. In consumer brands, many companies that have built up a successful brand name have blundered into such a duel by trying to transfer that brand onto something new. Mathematicians wouldn't blame them for trying, on the grounds that fixed points don't have to be unique: Zero squared equals zero, and one squared equals one. What's the problem? The problem is that the public will be confused. Heinz, for example, began life in 1869 as America's original pickle company, but before the 19th century was through it added ketchup to its product mix, unaware of the implied risk for a company whose name "meant" pickle. The good news was that Heinz was successful in this new venture, so much so that today their name is synonymous with ketchup. But the company couldn't avoid what marketing experts Jack Trout and Al Ries call the Teeter-Totter Principle, which accurately predicted that they would lose their leadership in the pickle market. And so it was: Vlasic, not Heinz, has been the leading pickle company for a generation.

Another source of dueling fixed points is the ever-changing identities of political parties. For example, when presidents break with their own party on a policy position, their supporters might be forced to decide whether their allegiance belongs with the individual or with the party. Duels can also spring unexpectedly from within a party platform. For example, "family values" has the sound of a bedrock political fixed point, promoting marriage, the nuclear family, and all sorts of good stuff while excluding such unwanted intrusions as unwed motherhood. But a pro-life position must ultimately embrace unwed motherhood; thus, family values weren't fixed in the first place, despite their obvious allure.

I confess that most of the generational fixed points we've seen have been a wee bit depressing for the sentimentalists among us. How nice it would be if the hardware store in the town we grew up in were still there when we returned home 30 years later. But no, it's now a Walgreen's. Fixed points are guaranteed only in closed, bounded environments, and the passage of time knows no boundaries. Yet there's hope: If some of our cherished memories weren't fixed the way we thought they were, perhaps some of humankind's biggest arguments and scourges won't be around forever, either. So, if you're tired of the way abortion dominates domestic politics, remember that it wasn't always so and needn't always be so in the future. Even the presumption that the Middle East will forever be wartorn isn't an absolute truth. Admittedly, it doesn't help us much if these problems take a millennium to solve, but recognizing the potential for change is a vital ingredient in the change itself.

And even if accepting the vagaries of change isn't fun, the philosophizing that can help us through is already there. "Every new beginning comes from some other beginning's end." Thus spoke Semisonic in their smash 1999 hit "Closing Time." In other words, the hardware store that we were sentimental about could have been decried by the generation that preceded us. And the generation that follows could be sentimental when the Walgreen's gets replaced by something else. Naturally I don't admit to anyone that my attitudes toward life were changed by a song lyric. Fortunately, I don't have to. Semisonic was just quoting the Roman philosopher Seneca, who uttered those words during the first century AD. Closing time indeed.

TWELVE
THE FULL MONTY HALL

Television game shows are almost by definition structured as puzzles. *Concentration* was a rebus puzzle. *Wheel of Fortune* is a bunch of word puzzles. But it is possible for a puzzle to make its way into the lore of a game show even though it isn't part of the original design.

Our tale begins with *Twenty-One*, one of the tainted big-money question-and-answer programs of the late 1950s. *Twenty-One* is now best remembered as the setting for the 1994 movie *Quiz Show*, which took a director's lens to the scandal in which quiz show producers coached the contestants and fed them answers in advance. The specific focus of the movie was the battle of wits between 29-year-old G.I. college student Herbert Stempel and Columbia professor/literary scion Charles van Doren. In one bizarre sequence, understandably bypassed by *Quiz Show*, Stempel was asked by host Jack Barry to link the four great voyages of Columbus with the following four destinations: the Virgin Islands, Santa Lucia, Hispaniola (Haiti and the Dominican Republic), and South America.

This question came with unusually high stakes, as van Doren was already sitting on 21 points and Stempel needed to ace his multipart question to tie the game. He started smoothly. He had no trouble associating Hispaniola with Columbus's first voyage and South America with his third. (You wouldn't have any trouble either if you had received the answers beforehand.) After a bit more manufactured thought and discomfiture he came up with Santa Lucia for voyage number four. And then, on the brink of success, he inexplicably stopped—for 12 long seconds. An eternity for commercial television, and plenty of time for a puzzler's memory banks to dredge up the following classic:

> **Seven businessmen go to a meeting over dinner at a local restaurant. Each checks his coat before entering the dining area. When the men finally leave the restaurant, their coats are returned to them randomly. What is the probability that exactly six men receive their own coats?**

People have been known to recoil from the phrase "what is the probability?" as if the words actually mean "You stupid fool, you don't have a prayer of solving this puzzle." Agatha Christie's Dr. Haydock chimed in on this subject in "The Mirror Crack'd from Side to Side": "I can see looming ahead one of those terrible exercises in probability where six men have white hats and six men have black hats and you have to work it out by mathematics how likely it is that the hats will get mixed up and in what proportion.

If you start thinking about things like that, you would go round the bend. Let me assure you of that!"

This time around, Dr. Haydock would have done better to listen closely before flying off the deep end. After all, if six men receive their own coats, the same must be true of the seventh man, so the probability of exactly six matches is zero. *Exactly* is the key word here, and the puzzle in its written form is made much easier when that word is italicized in the first place. The important issue is that the businessmen-and-coats puzzle isn't a probability puzzle at all, which is just as well because, as we will soon find out, probability is an area where few of us can claim any natural aptitude.

For the moment, however, let's get back to Herbert Stempel, still transfixed in his isolation booth in front of what must have been a bewildered studio and television audience. Finally betraying some emotion, Stempel lets out an "oh what a ninny I've been" kind of chuckle before telling Jack Barry what Barry already knew—and what Barry knew Stempel was instructed to say—that Columbus must therefore have visited the Virgin Islands on his second voyage. Well, of course he did. The Virgin Islands and the second voyage were guaranteed to match up, just like the seventh businessman and his coat.

Moviegoers with long memories might be surprised to hear that the Columbus charade came immediately after Stempel failed to identify *Marty* as the 1955 Academy Award winner for Best Picture. In *Quiz Show*, Stempel was eliminated altogether by the *Marty* question. That wasn't the only occasion in which *Quiz Show* took some convenient but harmless liberty with the facts on the ground. For our purposes, the most noteworthy truth-bending was that when the scandal finally exploded in the summer of 1958, the real Jack Barry, unlike his *Quiz Show* counterpart, was taking a few weeks off from his hosting gig. In his place was an up-and-coming Canadian announcer named Monty Hall.

Monty Hall was not a household name in 1958. When the demise of *Twenty-One* and its ilk brought the bull market for quiz-masters to an abrupt halt, he spent a year doing color commentary for the New York Rangers and then took over for Jack Narz as the host of a lighthearted effort called *Video Village*, a conceptual fore-runner to Stubby Kaye's *Shenanigans*. Hall's career-changing mo-ment wouldn't arrive until 1963, when he and his business partner, Stefan Hatos, launched an NBC daytime game show called *Let's Make a Deal*.

The thrust behind the show was not large cash payouts, still rendered toxic by the quiz-show scandal's enduring shadow. In-stead, the idea was that Monty would stroll through the studio audience and choose contestants who would be given a chance to win and trade prizes of great value or maybe no value at all. So instead of Herbert Stempel agreeing to "risk" $29,000 for the op-portunity to win $50,000, a *Let's Make a Deal* contestant might trade a comb or makeup case for whatever was behind an onstage curtain, soon to be revealed by Monty's iconic assistant Carol Merrill. Any chance of winning the Ford Mustang behind curtain number 2 was offset by the chance of winning a so-called zonk— either the truckload of spoiled cottage cheese behind curtain num-ber 1 or the live goat behind curtain number 3.

You could say that whereas *Twenty-One* rewarded contestants in a clandestine fashion for their telegenicity, *Let's Make a Deal* rewarded them openly for their creativity. In the early episodes of *Let's Make a Deal*, the studio audience was attired in the normal street clothes of the era: coats and ties for the gents, and conserva-tive suits and dresses for the ladies. The dress code changed in early 1964, when Monty Hall came across a woman carrying a sign reading "Roses are red / Violets are blue / I came here / To deal with you," and he selected her to be a contestant. Shortly afterward, he chose someone wearing a clownish hat, and thereafter wacky costumes became standard fare. What followed was an extraordinary success, and it was easy to imagine that the name Monty Hall would be forever linked with the show. Not quite. Instead, his name is forever linked with a puzzle.

> **Suppose you're on a game show, and you're given the choice of three doors: Behind one door is a car; behind the others, goats. You pick a door, say number 1, and the host, who knows what's behind the doors, opens another door, say number 3, revealing a goat. He then says to you, "Do you want to pick door number 2?" Is it to your advantage to switch your choice?**

That's the way the Monty Hall problem appeared in Marilyn vos Savant's "Ask Marilyn" column for *Parade* magazine on September 9, 1990. The problem was submitted by a reader named Craig Whitaker, and Whitaker's problem was in turn based on a 1975 letter sent by Steve Selvin of Berkeley's School of Public Health to the editor of *American Statistician* magazine. *Parade* is, of course, more widely read than *American Statistician*, and Marilyn's column placed the Monty Hall problem squarely in the public eye. With the standard disclaimer that the problem as posed above wasn't an option that Monty Hall actually offered to *Let's Make a Deal* contestants, the question remains: Do you switch your original choice of doors or don't you?

Perhaps this is a good time to recall the suggestion of a few pages ago—that we as human beings just aren't very good with probabilities. The Monty Hall Paradox, as it has come to be known, pitted mathematician against mathematician, brother against brother, and generated more reader mail than anything else that has ever appeared in *Parade*. Even Paul Erdös, one of the most prolific mathematicians of the 20th century, was said to have been fooled by the problem. As Stephen Jay Gould said in *Dinosaur in a Haystack*, one of his many essay compendiums, "Misunderstanding of probability may be the greatest of all impediments to scientific literacy."

Marilyn's claim, 100 percent correct, is that you should switch from door number 1 to door number 2. But readers didn't buy it, and that's when the mail started flooding in. The logic of Marilyn's protestors took various forms, but the most common theme was that once Monty Hall showed the goat behind door number 3, only two choices remained, so the odds of either door containing the prize had to be 50 percent. No need to switch.

In general, the concept of a 50-50 chance seems like the simplest thing you could imagine, as in heads versus tails. But there are plenty of ways to go wrong, and two separate classes of mistakes emerge. The first class would be the almost 50-50s. In writing the quantitative reasoning manual *What the Numbers Say*, my coauthor David Boyum and I went an honest 50-50 and each contributed one such example—that is, a not-quite-50-50—to the final manuscript. His was that the odds of a newborn being male are 106 to 100, slightly greater than the expected 50-50. My own example was the stock market, where it's common to think of up days and down days as if patterned by the flip of a coin, but the likelihood of an up day turns out to be roughly 52 percent, not 50 percent, according to a century's worth of trading sessions.

But it's also possible to be off by a mile in a "50-50" situation.

Management consultant Bill Givens used to say that in any situation where we think we have a 50-50 shot—usually a matter of choosing between left and right, whether looking for a light switch in an unfamiliar room or vacillating at a stoplight in an unfamiliar town—we almost never get it right. Givens's idea was formalized by Andy Rooney in the form of the 50-50-90 Rule: "Anytime you have a 50-50 chance of getting something right, there's a 90 percent probability you'll get it wrong." The underlying fallacy behind this rule is that we tend not to remember the occasions when we guess correctly, but when we guess wrong and suffer an inconvenience as a result, our error sticks with us.

Deviations from 50-50 can be sneaky. Basal cell carcinomas (benign skin cancers) are usually found on the face, but on which side? Again the issue is left versus right, but the distribution is far from 50-50. Men get them predominantly on the left side and women (especially older women) on the right, coinciding with their exposure to the sun when a man drives and a woman sits in the passenger seat. One supposes that in the UK it's the other way around. And Italian researchers determined that requests made to someone's right ear are twice as successful as requests made to the left ear, apparently because of the asymmetric fashion in which the brain receives auditory signals. (The underlying context was asking for a cigarette in a crowded disco, a bygone act in the United States but one in which the very act of approaching an ear—either ear—wasn't deemed weird in the first place.)

As for Monty Hall and the question of switching doors versus not, the apparent 50-50 choice induced readers to think that a switch wasn't mandated. But there's a major flaw with that line of thinking: If the odds really were 50-50 once door number 3 was revealed, then Monty's actions somehow raised the odds that the original choice was correct. Yet given that our fictional Monty didn't do anything beyond what you knew he could do, how could those odds change? The subtle brilliance of the Monty Hall puzzle's wording is that when it poses the question of whether you should switch, it sounds as though it must be a close call, an "almost" 50-50, as it were. But this one clearly belongs in the other category. Not only should you switch, you *double* your chances of winning if you do so, from 1 in 3 to 2 in 3. The 50-50 shot is never in the picture.

Let's look at the same issue in a slightly different way. Whatever door you happened to choose—we'll say it was door number 1—your original chance of being correct was 1 in 3. That's beyond dispute. The chance that the prize is behind either door number 2 or door number 3 is therefore two-thirds. Once Monty shows the zonk behind door number 3, that door is ruled out, and door number 2 must capture all of that 2 in 3 chance, so you should switch to it. And that's what's so strange about the Monty Hall Paradox. As bad as we humans are at probabilities, the idea that our original likelihood of winning was 1 in 3 is the one thing that is pretty damn clear. Yet to be fooled by the paradox we must discard that very clarity.

The usual way to convince nonbelievers that Marilyn was correct is to imagine not three but 100 doors. In this enhanced version of the puzzle, you choose a door and Monty proceeds to open 98 of the remaining 99 doors, each of the 98 revealing a goat. *Now* would you switch your choice? Of course you would. It's obvious, isn't it? (This aspect of the Monty Hall problem is unusual in that an important insight for a puzzle involving a very small number—three doors—is obtained by looking at a large number instead. Elsewhere in the puzzle world, it's much more common to investigate small numbers to obtain insights about larger ones.)

Another way to dispel the prevailing misconceptions about the Monty Hall problem is to show what an actual 50-50 puzzle looks like. One such example revolves around two sealed envelopes that have money inside them. The exact amounts are not specified, but one envelope is known to contain twice as much as the other. You pick one envelope and open it to discover that it contains $100. Do you switch to the other one? If the question sounds silly to you, that's a good sign, because it *is* silly, as long as you don't overthink the problem. The trap that awaits overthinkers is that the other envelope could have $200 in it, and it could have $50. The average of these two numbers is $125, which is higher than the $100 with which you started. The $25 premium makes switching seem attractive, but that's absurd. If ever there were a straight 50-50 shot, this problem provides it, and of course it is entirely different from the Monty Hall problem.

A final way to end the mental stalemate surrounding Monty Hall is via computer simulation. It is said that Paul Erdös adopted a pro-Marilyn stance only after a colleague arranged for a so-called Monte Carlo simulation involving hundreds of trials of the basic Monty Hall setup. When these came out 2 to 1 in favor of switching, Erdös conceded the point.

The concept of computer trials was extended in a subsequent experiment conducted by Walter T. Herbranson and Julia Schroeder of Whitman College in Walla Walla, Washington. They simulated the Monty Hall problem using three keys, one of which led to food for the participants. Herbranson and Schroeder had the ability to deactivate one of the keys once the initial choice was made and before a second choice was offered (that is, stay or switch), thereby mimicking the Monty Hall paradigm as best they could. The results showed that dramatic improvement was possible. On the first day, responders stayed with their original choices

64 percent of the time. By the 30th day of the experiment, though, participants *switched* 96 percent of the time. These results far surpassed other laboratory replications of the Monty Hall problem, but whether that's good news or bad news is something of a philosophical question. The complication is that in this case the subjects were not humans, whose probabilistic intuitions we have cast doubt on. No, the participants in this particular experiment were six silver king pigeons. Their reward was a bunch of mixed grains.

The Monty Hall Paradox has shown up in animal experiments even when it wasn't invited. In 2007, researchers at Yale sought to explore the roots of cognitive dissonance—the term applied to circumstances in which the mind is forced to process two conflicting thoughts at the same time. Cognitive dissonance inevitably results in rationalization, as in Aesop's fable of the fox and the grapes, in which the grapes looked luscious but were deemed sour once they proved unreachable. The Yale researchers were trying to see if cognitive dissonance applied to real animals, and capuchin monkeys were the lab animals of choice.

The experiment began by giving the monkeys a choice of different colored M&Ms. The monkeys that exhibited no particular preference among red, blue, and green were then given only two choices, red and blue. When the red M&Ms won out in this direct test, the monkeys were given a choice between blue and green. The fact that green won out over blue in this second test was taken as evidence that the monkeys had rationalized their prior choice. This type of negative momentum is a well-documented consumer phenomenon. According to the 1981 marketing bible *Positioning*, "In a small corner of the brain is a penalty box called 'loser.' Once your product is sent there, the game is over."

But colors and experiments have a tricky history. A letter

written by the prominent British statistician George Udny Yule (1871–1951) contained the following cautionary passage:

> Isn't it extraordinary how difficult it is to get a sample really random? Every possible precaution, as it may seem, sometimes fails to protect one. I remember Greenwood telling me that, in some experiments done by drawing different coloured counters from a bag, there seemed to be a bias against one particular colour. On testing, they concluded that this colour had given the counters a slightly greasy surface, so that it tended to escape the sampler's fingers.

For that matter, colors and M&Ms have a tricky history all their own. Red M&Ms were pulled off the market from 1976 and 1987 even though they didn't contain the carcinogenic red dye that everyone was worried about, green M&Ms have been rumored to be an aphrodisiac, and in 2009 the dye for blue M&Ms was shown to reduce the effect of spinal cord injuries in rats.

The monkeys in the Yale experiment weren't biased by these various details, but the entire effort faced an unexpected complication. As economist M. Keith Chen, also of Yale, pointed out, there was a possible explanation for the monkeys' behavior that undercut the cognitive dissonance theory. If the monkeys had a built-in preference, however slight, for red versus blue, their complete preferences regarding the three colors would take one of the three forms shown in Figure 12-1, where an arrow denotes preference.

$$RED \rightarrow GREEN \rightarrow BLUE$$
$$GREEN \rightarrow RED \rightarrow BLUE$$
$$RED \rightarrow BLUE \rightarrow GREEN$$

FIGURE 12-1

At this point, the Monty Hall arithmetic intrudes on the proceedings. The monkeys have a preference for green over blue in two of the three cases, even if they weren't trying to rationalize their prior choice. According to Chen, a similar problem besets virtually all laboratory representations of cognitive dissonance since the theory first took hold in 1957.

The Monty Hall Paradox has applications in many probability-based games, especially card games. In bridge, it is considered axiomatic that skilled declarers must know how to play the percentages, but the theoretically correct way to handle a certain card combination may not end up being correct toward the end of the hand, when additional information is available about the defenders' distributions.

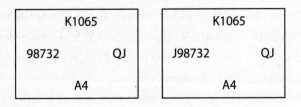

FIGURE 12-2

Perhaps the best-known probability conundrum in bridge is the principle of restricted choice. This principle isn't quite an everyday event, but it applies with special force when you as declarer have a holding of, say, the ace of diamonds in your own hand with the king and the 10 in dummy. When you play the ace, the queen drops on your right, giving you an unexpected chance at three tricks in the suit. Do you play toward your king, hoping that your opponent was originally dealt just the queen and the jack (as in the left-hand diagram of Figure 12-2), or do you play to the 10, hoping that the queen was singleton (as on the right)? Absent any other compelling information, the principle of restricted choice says you should play the 10. The rationale is that your opponent, if holding the doubleton queen-jack, could have played either card because they were of equal value. The queen will therefore appear 100 percent of the time when it is a singleton and only 50 percent of the time when it is a doubleton, so you should go ahead and take the finesse with the 10. If Marilyn vos Savant wants to take up restricted choice, she is practically guaranteed of another flood of reader mail.

The final tribute to the Monty Hall Paradox is that it has received far more attention than any other thought-provoking scheme from the game-show world. Not long after *Let's Make a Deal* hit the airwaves, NBC came out with a game show called *I'll Bet* (later reincarnated as *It's Your Bet*). In the show, one member of a couple would be given a question and would have a chance to bet whether his or her spouse would be able to answer it. The amounts were between 0 and $100.

But a funny thing happened on *I'll Bet*. Consistently, you'd see a contestant say, "I'll bet $100 that . . ." and then take several seconds before deciding whether the wager should be on yea or nay. Such vacillation makes sense when coupled with small bets, because both are indications of uncertainty about the partner's knowledge of the subject. But why would vacillation accompany large bets? The answer, presumably, is that the opportunities offered by *I'll Bet* didn't arise every day. It's not as if contestants could count on dozens of appearances and play the odds accordingly. They felt a pressure to score with the material presented them, and the result was this disconnect between the confidence implied by their wager and the uncertainty conveyed by their hesitation. To make matters worse, the yes-no decision was made by moving a right-handed lever, while the amount of the bet was set with the left hand, so the game literally depicted the right hand not knowing what the left hand was doing.

Unfortunately, this situation occurs all too often in the real world. Inexperienced investors tend to agonize over the question of whether to buy a stock but spend quite a bit less time exploring the question of how much they should purchase. The principle should be that the less confidence you have in a particular company, the smaller your position, with zero being the ultimate expression of no confidence. But many investors simplify the process by wrongly turning it into a yes-no decision (with a high dollar figure) rather than a continuum of choices.

Strategies for *The Price Is Right* have a different type of real-world application. The basic objective of the show's contestants is to come as close as possible to guessing the price of an object without going over. If your opponent makes a guess that is clearly too low, your best counterstrategy is to simply bet $1 more—not because it necessarily represents your best estimate at the object's value but because your objective is to win the game. As in the apocryphal tale of the two hikers who encountered a bear, their objective wasn't to outrun the bear; it was to outrun the other guy.

This type of strategy is not widely available in real life outside of auction houses, but it is occasionally exploited by second husbands. The game of love theoretically involves hundreds of potential adversaries, not just one, but if you marry a woman whose first husband was a real lout, you don't have to be perfect; you just have to be demonstrably superior to him, or so the theory goes. If he was a drunk, you don't need to be a teetotaler; you just need to drink less. If he was a skinflint, you don't need to be a spendthrift;

you just need to be less cheap. And so on. The bar has been set low, and your objective is to sail over it, not trip yourself up by raising it too high.

As you can see, other game shows come up far short of the Monty Hall standard, but we would never close on such a down note, especially when the world of romance can be invoked for a more upbeat signoff that is a puzzle in its own right. It's a puzzle I won't try to solve, either because there is no answer or because your answer is better than mine. Maybe it's just food for thought. Recall that when Monty selected one of the two unchosen doors, his choice meant something, even though it had to be one or the other. Well, when a couple finds out that they are having a child, they might not know from the outset whether it will be a boy or a girl, but they do know that it will be one or the other. But gender discovery isn't like a game show with a car and a goat. Nor is it like a sporting event, in which either Team A wins or Team B wins and when the event actually arrives, one group is happy and the other unhappy. As we know, when the gender of a baby is finally identified, the announcement "It's a boy!" generates great excitement among those who receive the message . . . while the announcement "It's a girl!" also generates great excitement among the very same people. Given that the child had to be one or the other, how can this be?

ACKNOWLEDGMENTS

It's time to admit that I didn't write this book alone. Let me start by thanking my agent, Jennifer Griffin, for her tireless work on my behalf. She hooked me up once again with Marian Lizzi, my editor at Perigee for the second time, who was still as enthusiastic as ever. It was also a pleasure to again work with Christina Lundy and the rest of the Perigee team. Candace B. Levy did a great job with the copyediting and taught me more than a few wrinkles along the way.

Norton Starr did me the tremendous service of reading a few preliminary chapters and righting my course. I also called upon Rob Tittman and Mitchell Stokes to chime in when I stumbled into issues that could tap their clinical expertise. Tamsin German was helpful in explaining some of her academic work, especially her collaboration with Greta Defeyter. Charisse Nixon was also a useful resource. I don't think I specifically called on David Boyum during the preparation of this manuscript, but I can point to several passages that were influenced by his thinking, and the book is much the better for it.

Book research can take you to surprising places and can trigger old memories. This particular book gave me a chance to reconnect with Peter Hodgson, who didn't recall the story that prompted me to call him in the first place but gave me several other great stories and viewpoints in return. Walter Reed at Emory University was a similar long-lost connection, and all I can say is that I'm lucky I know people who are so well read, articulate, and willing to help.

Raymond Smullyan, Jerry Slocum, Greg Ross, Lynne Emmons, Chris Holmes, Peter Costa, Tom Rodgers, Pam Miller, Timothy Smiley, Jonathan Levav, and the Image Zoo all helped me with images, permissions, choice quotes, or a combination of the above. Finally, I'd like to thank my puzzle mentors and supporters, namely Will Shortz and the late Eugene T. Maleska in the area of crosswords; Peter Gordon at Sterling Publishing; and Bill Ritchie, Andrea Barthello, Tanya Thompson, Sarah Hart, and the rest of the great crew at ThinkFun, always my first call when I get a new game or puzzle idea.

ABOUT THE AUTHOR

Derrick Niederman has maintained a professional and recreational interest in puzzles for many years. He received a BA in mathematics from Yale in 1976 and a PhD in mathematics from MIT in 1981. He spent almost 20 years in the investment business, both as an analyst and as a columnist, and in the last decade has written a dozen books in the areas of finance, mathematics, and puzzles. He has also been a regular contributor of crossword puzzles to the Sunday *New York Times* since 1981.

In 2008, his geometric puzzle 36 Cube was launched by ThinkFun, a leader in educational games and puzzles. In 2011, ThinkFun introduced a second puzzle from the author's collection, a hybrid of Word Search and Tetris called PathWords.

Dr. Niederman lives in Charleston, South Carolina, and teaches mathematics at the College of Charleston.